The Never-Ending Story of Life

Carlos E. Semino

The Never-Ending Story of Life

A Brief Journey through Concepts of Biology

Carlos E. Semino
Tissue Engineering Research Laboratory
Department of Bioengineering
IQS-School of Engineering
Ramon Llull University
Barcelona, Spain

ISBN 978-3-030-75971-1   ISBN 978-3-030-75969-8   (eBook)
https://doi.org/10.1007/978-3-030-75969-8

This Springer imprint is published by the registered company Springer Nature Switzerland AG.
The registered company address is: Gewerbestrasse 11, 6330 Cham, Switzerland

To my beloved children
Santiago and Alba

Preface

For humankind, the most irreducible idea is the concept of life itself. Life is the principal driving force behind many disciplines, including philosophy, literature, biological sciences, medicine, mythology, and theology, but no matter how much we know about it, it will always remain our greatest mystery. Although some scientists study the origins of life to examine the group of pre-biotical, biochemical reactions that may have spontaneously created it, this book begins from the premise that life is a fact—that it is everywhere; that it takes infinite forms; and, most importantly, that it is intrinsically self-perpetuating. Rather than exploring how the first living forms emerged in our universe *"The never-ending story life"* begins with our first primordial cell ancestor and tells the story of life's journey—how it began when that first cell diversified into many other cell types and organisms, and how it has continued until the present day.

This small book is designed to explain complex ideas in biology in a simple way, but not simplistically, with a special emphasis in taking care of the language as well as illustrations that work for all possible readers. Thus, the central concept that this compact but essential book illuminates for biology students and nonbiologists alike is the fact that life is essentially an infinite process, transmitted from generation to generation. Curiously, the strategy that life uses to "survive" is based on a simple mechanism: it generates a vast amount of highly diverse organisms so that every ecological niche is inhabited by specialized living forms. The remarkable heterogeneity of these organisms ensures that a great proportion of them will be able to adapt to the constantly changing environment. Through this process, life can take infinite biological forms, from simple cells to complex animals, fungi, and plants. Each of these individual organisms has a lifespan and must generate descendants (i.e., offspring) that will be different from their ancestors, because living organisms have intrinsic mechanisms to generate genetic diversity. This diversity is generated when a cell divides to produce two new ones, because genetic changes (i.e., mutations) occur independently during the division process, producing daughter cells that are not exactly identical to the original cell. During this process, life transmits itself to the next generation simply by splitting into two and "passing" from one cell into the two new ones.

Thus, the book explains how life is passed on during every cell division. It never "jumps" from one organism to another; instead, it "flows" on to the next generation. This is a very important concept, because regardless of whether an organism is unicellular or multicellular—whether it is a small bacterium or a whale—life flows exclusively through the process of cell division. This is why cells are the most evolutionary form of life and the smallest building block capable of passing it on. This may be hard to understand at first, but when we consider that even the most complex animals and the largest trees almost all rely on unicellular cells (gametes) to transfer life to the next generation, it becomes clear that cell division is the essential mechanism of life's self-perpetuation. This is also true in some cases where complex structures—pieces of tissues rather than single cells—are used to generate more organisms, as in the processes of animal fragmentation and vegetative propagation in plants, where life splits up to continue self-propagating.

In all of these processes, life is immortal, but organisms are not. Each organism has a lifespan—hours, days, weeks, or years, depending on the case—but in the end, it must die. The reason for this is simple: organisms have to generate diversity, because this is the only natural way for it to create genetically distinct descendants, and in most multicellular organisms, this process naturally leads to death. Cell division drives growth and development, maintains tissue and organ homeostasis, executes the body's regeneration and repair, and generates gametes, but it also promotes aging and death. In other words, each organism has an intrinsic program that eventually leads to death. Thus, death is an essential part of life, and each individual organism is a "carrier" of life that can pass it on to the next generation before dying, thereby continuing with the never-ending story of life.

I would like to thank the many people who have contributed to the organization, editing, revision, drawings, and ideas in this book. In particular, I extend my most sincere thanks to Nausika Betriu, Claire Jarrosson Moral, Alba Costales, Oriol Quintana Rubio, Gustavo Tiscornia, Mariana Resnicoff, Joel Paul, and John Shakespear.

Barcelona, Spain   Carlos E. Semino
May 2021

Contents

Abbreviations

AIUs	Artificial intelligence containing units
Ar	Argon
ATP	Adenosine 5-triphosphate
BA	Basal membrane
BYA	Billion years ago
BY	Billion years
Ca^{2+}	Calcium
CO_2	Carbon dioxide
CV	Central vein
CLP	Common lymphoid progenitor cell
CMP	Common myeloid progenitor cell
DNA	Deoxyribonucleotide acid
ESCs	Embryonic stem cells
ESI	Earth similitude index
ER	Endoplasmic reticulum
ECM	Extracellular matrix
Epi	Epiblast
ExE	Extraembryonic ectoderm
GMP	Granulocyte monocyte progenitor cell
H_2O	Water
HSCs	Hematopoietic stem cells
HA	Hepatic artery
HERVs	Human-derived retroviruses
ICM	Inner cell mass
ISCs	Interstitial stem cells
IVF	In vitro fertilization
LPCs	Liver progenitor cells
LT-HSCs	Long-term hematopoietic stem cell

LY	Light-years
MEP	Megakaryocyte erythrocyte progenitor cell
MSCs	Multipotent stem cells
MYA	Million years ago
NASA	National Aeronautics and Space Administration
N_2	Molecular nitrogen
O_2	Molecular oxygen
PV	Portal vein
Pro-B	Pro-lymphocytes B
Pro-T	Pro-lymphocytes T
SCs	Stem cells
ST-HSCs	Short-term hematopoietic stem cell
TE	Trophectoderm
TERT	Telomerase reverse transcriptase
TR	Telomerase RNA
RNA	Ribonucleotide acid
UV	Ultraviolet

1

The Origin of Eukaryotic Cells and Multicellular Organisms

Summary First signs of life on Earth were most probably in the form of unicellular organisms that rapidly diverged into different types while adapting to all conceivable environments. From those, the first eukaryotic cells appeared giving rise to the origin of multicellular organisms that after an unimaginable series of changes and adaptations ended in what we know today as fungi, plants, and animals. This chapter explains how this long and complex process most probably happened in the last billions of years since the origin of life.

From the moment the invaders arrived, breathed our air, ate and drank, they were doomed. They were undone, destroyed, after all of man's weapons and devices had failed, by the tiniest creatures that God in his wisdom put upon this earth. By the toll of a billion deaths, man had earned his immunity, his right to survive among this planet's infinite organisms. And that right is ours against all challenges.
For neither do men live nor die in vain.
—H. G. Wells, *The War of the Worlds*

First Steps of Life on Earth

We do not really know how the first cells were created, but we do know that 3.7 billion years ago (BYA), the first unicellular organisms were already populating our planet. This is why whether these organisms developed here on Earth, came from outer space, or formed as a hybrid between terrestrial and extraterrestrial organisms, we call this original cell or common cell the "ancestor."

Now, the study of the prehistory of cells is a difficult job, because we do not have physical registries for cells in the way paleontologists have registries for fossils. Interestingly, the evolution of cells is therefore studied by comparing the differences among the genetic material of organisms that are alive today. Why? Because genomes—the genetic material that is passed down from generation to generation, which is composed in most organisms of deoxyribonucleic acid (DNA)—contain not only information about how each particular organism will be able to build its cells, tissues, organs, body shapes, etc. but also a record of the different modifications (mutations) that an organism has experienced over time. The information we obtain from comparing genomes is so powerful that it can take us almost all the way back to the origin of cellular evolution. To put it simply, more closely related organisms have more similar genomes than organisms that are less closely related. By comparing this genetic material, it is possible not only to classify all the cell types that presently exist but also to predict all their potential ancestors. Thus, from looking at cells' family trees, we now have a good idea of how they have evolved over the last 3.7 BY and of their almost infinite capacity to generate diversity.

These first primitive, unicellular organisms adapted to colonize every single niche in the surface of the Earth. It is believed that the first primitive cellular common ancestor evolved into two main cell lineages: early bacteria and archaea[1]-proto-eukaryotic[2] cells. Both cell types were composed of a cellular membrane and a circular DNA molecule that contained their genetic material. Bacteria have continued evolving independently ever since into hundreds of different species with innumerable adaptations. Crucially, very early on, a group of bacteria called cyanobacteria started photosynthesizing—using light energy to reduce carbon dioxide (CO_2) into carbon-containing molecules, like carbohydrates—and producing molecular oxygen (O_2) as residue. At that time, oxygen was toxic for most existing anaerobic bacteria. Over the course of 1 BY, O_2 accumulated in the atmosphere until it became so prevalent that anaerobiosis was only possible in special niches, causing a drastic reduction in the number of anaerobic species on Earth. Interestingly, by 2.5 BYA, O_2 was used by aerobic heterotrophic bacteria, which expanded very quickly. These new bacteria were capable of oxidizing reduced carbon-containing molecules to produce highly energetic molecules like adenosine-5-triphosphate (ATP),

---

[1] Archaea is a class of microorganism without nucleus or organelles that constitutes one of the three distinct evolutionary lineages of modern-day organisms. It is also called *archaebacteria* and *Archaeans*.

[2] Eukaryote is a class of unicellular or multicellular organisms in which the cells contain membrane-limited nucleus and membrane-limited organelles including endoplasmic reticulum (ER), mitochondrion (animal and fungi), and chloroplast (vegetal), which constitutes one of the three distinct evolutionary lineages of modern-day organisms; it is also called *eukarya*.

generating CO_2 and water (H_2O). In this way, phototrophic and heterotrophic bacteria conquered practically the entire planet; today, they constitute what is called the prokaryotes.

In the other lineage that evolved from those first unicellular organisms, archaea-proto-eukaryotic cells diverged and gave rise to archaea and eukaryotic cells (Fig. 1.1, *The origin of eukaryotic cells*). Archaea have continued evolving, developing into their own group, but the eukaryotic group's

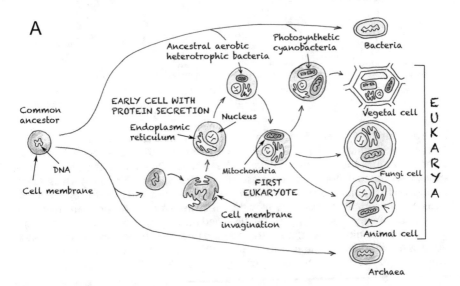

Fig. 1.1 The origin of eukaryotic cells. (a) A common ancestor cell appeared on Earth about 3.7 billion years ago (BYA). It initially diverged into bacteria (which later became its own kingdom) and archaea-proto-eukaryotic cells. Archaea-proto-eukaryotes diverged into archaea (which also formed its own kingdom) and early eukaryotic cells. These latter cells underwent cell membrane invagination to form the nucleus (which contains their DNA) and the endoplasmic reticulum (which provides the capacity to synthesize proteins that can be secreted). Then, an endosymbiotic event took place: ancestral aerobic heterotrophic bacteria entered into a eukaryotic cell ancestor, which later turned into mitochondria. This marked the formation of the first eukaryotic cell, which diverged into today's fungi and animal cells. In a second endosymbiotic event, ancestral aerobic photosynthetic bacteria (cyanobacterium) entered into the early eukaryotic cell, which later turned into a chloroplast. This new cell type became the origin of vegetal cells. Fungi, animal, and vegetal cells all belong to the Eukarya Kingdom. (b) An early common ancestor with RNA-based genome appeared on Earth about 3.7 BYA. From this ancestor, three basic types of cell domains emerged, a common bacteria-archaea ancestor with a DNA-based genome that later diverged into bacteria and archaea, respectively, and, a third one, a proto-eukaryotic cell (RNA-based genome). This proto-eukaryotic cell presented an active cytoskeleton and an invaginated cell membrane that formed an early endoplasmic reticulum, which provides the

(continued)

B

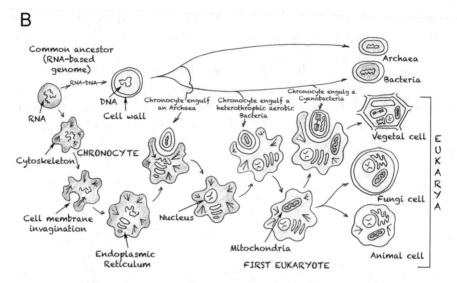

Fig. 1.1 (continued)   capacity to synthesize proteins that can be, eventually, secreted. This proto-eukaryotic cell is called chronocyte after engulfing an archaea. The achaea turned later into the nucleus, containing the achaea's DNA and, later, the retrotranscribed RNA-based genome that progressively was transported into the nucleus. After a second endosymbiotic event, chronocyte engulfed ancestral heterotrophic aerobic bacteria, which later turned into mitochondria, generating the first eukaryotic cell. Same as above, the first eukaryotic cell diverged into today's fungi and animal cells. In a third endosymbiotic event, ancestral aerobic photosynthetic bacteria (cyanobacterium) were engulfed by the primitive eukaryotic cell. The cyanobacteria later turned into chloroplasts. This was the origin of vegetal cells

evolutionary path was not so straightforward. Several theories have been postulated to explain the evolution of eukaryotic cells, and the most plausible ones have been selected and described here. The first currently used theory suggests that the early eukaryote cell was generated after a series of events that occurred about 1.5 BYA, including cell membrane invaginations that created specialized internal endomembrane systems such as the nucleus, which contained the cell's DNA, and the endoplasmic reticulum, which is a specialized organelle localized around the nucleus in charge of synthesizing proteins that could then be secreted to the exterior of the cell. Therefore, this was the origin of the first nucleated cell (Fig. 1.1a, *The origin of eukaryotic cells*).

Around that time, it is believed that the DNA molecule (which was circular in ancestor cells) became linear and associated with a special type of proteins, the histones (basic proteins with barrel-like structure used to roll up the DNA molecule to minimize it volume), which helped pack the DNA into a highly organized structure called chromatin, giving rise to the first eukaryotic chromosomes. In this way, very long pieces of linear DNA can be condensed, reduced in volume to fit into the cell, and placed together with other chromosomes. Essentially, this new eukaryotic cell type stored its genetic material (all the chromosomes) in a specialized organelle (nucleus) keeping it separate from the rest of the cell, the cytoplasm, where the main metabolic and synthetic activities take place. Remarkably, eukaryotic genetic material is in general very long. For instance, if all the chromosomes present in one single human cell are unpacked and linearized, they measure all together around 2 m long. In addition, if of all the linearized genetic material of all the cells present in a human of around 70 kg is measured (approximately 37 trillion cells or 37,000,000,000,000 cells), it will be 74,000,000,000 km long, which is equivalent to about 500 times the distance between the Earth and the Sun.

Moreover, if we do the same with other eukaryote cell types, for instance, a simple amoeba (a single-cell microorganism living in aquatic environments, which belong to the group of amoebae[3]), its genetic material will measure around 300 m, which is extremely large considering its simple lifestyle. Other examples like lilies, wheat, and salamanders are also very long (all around dozens of meters). This means that DNA packing by histones plays a very important role on eukaryotic cells because all this genetic material needs to fit into a very tiny nucleus of a cell of about 10 μm long, which is equivalent to 100 times smaller than a millimeter.

Then, this early nucleated cell underwent a process called endosymbiosis,[4] which means that ancestral aerobic heterotrophic bacteria were somehow incorporated into the eukaryote ancestor without having been digested. The bacterium therefore adapted to live inside of its new host and later evolved (maintaining its own original circular DNA) into what is known today as the mitochondrion, an organelle that is dedicated to produce energy in the form of ATP [1–3]. This new cell was the early eukaryote, which then evolved to form the ancestors of fungi and animal cells. Next, in a second endosymbiotic event, our early eukaryotic cell incorporated then an ancestral photosynthetic bacteria (cyanobacteria), which later evolved into what is called chloroplast (an organelle that has the capacity to use light to reduce CO_2 into carbon-containing molecules, like carbohydrates), creating the first vegetal cell. In this way, these three cell types (fungi, animal, and vegetal) evolved to form the

---

[3] Amoebae are a group of eukaryotic organisms, mainly unicellular, that do not have cell wall and therefore have the capacity of extending and retracting their cell body to eat or to move. Interestingly, amoebae can be found in many eukaryotic groups, including protozoa, fungi, algae, and animals.

[4] The endosymbiotic theory was first formulated in 1905 and 1910 by Russian botanist Konstantin Mereschkowski [1, 2]. In the 1960s, it was reinforced by microbiological evidences that Lynn Margulis found and presented officially in her book *Symbiosis in Cell Evolution* [3].

kingdom of unicellular and multicellular eukaryote organisms. Since mito-chondria and chloroplasts maintain its original circular DNA, they were therefore used by evolutionary biologists to compare their sequences and con-clude that they are indeed related to ancestral heterotrophic prokaryote and cyanobacteria, respectively.

Two alternative theories about the origin of the nucleus in early eukaryotic cells suggest that it was also produced after an endosymbiotic event. The first in proposing such idea was the Russian botanist named Konstantin Mereschkowski in 1905, who suggested that the nucleus evolved from bacteria engulfed by and entity named "amoebaplasm" which was not a bacterium [1]. Briefly, Mereschkowski sustained after looking at hundreds of species of lichens[5] that eukaryotic organelles, including the nucleus and the chloroplast, were the result of an intracellular symbiosis of bacteria with amoeba-like cell (or amoebaplasm). This idea was the fundamental principle of the previously described endosym-biont theory (or symbiogenesis theory) popularized by Lynn Margulis, who mainly pursued the idea that mitochondria and chloroplasts are the descendants of heterotrophic prokaryote and cyanobacteria, respectively.

The first theory in favor of endosymbiont origin of the nucleus claims that the early eukaryotic cells were the product of a fusion of an archaeon with a bacterium. In this model, the archaeon become the nucleus. This theory is symbolized as E = A + B or in words Eukarya = Archaea + Bacteria and is called the AB hypothesis. Then, the second alternative theory about the origin of the eukaryotic nucleus was postulated by Hyman Hartman, an evolution-ary biologist at MIT (Massachusetts Institute of Technology), who extended Mereschkowski's theory in favor of the nucleus being the product of an endo-symbiotic process but not by the fusion of an archaea and a bacterium (Fig. 1.1b, *The origin of eukaryotic cells*). The question Hartman asked was Who engulfed whom? In principle, Archaeans and bacteria cannot engulf themselves or other microorganisms since these cells are covered by a hard cell wall that protects them and, at the same time, makes them a very rigid organ-ism. Moreover, in order for a cell to engulf another cell, a complex structure named the cytoskeleton is required, which allows a soft and flexible cell to wrap itself around another smaller cell (a type of primitive amoeba-like cell). This primitive cell capable of engulfing Archaeans and bacteria was named chronocyte, which was probably a soft and flexible cell, without a cell wall and a functional cytoskeleton apparatus [4, 5]. Since only eukaryote cells present these characteristics today, this suggest that the chronocyte, which appears as

---

[5] Lichens are species of organisms that resulted from the symbiosis between a fungus and one or more algae.

an independent group, evolved into a eukaryotic cell by engulfing first a DNA-based Archaean (which subsequently formed the DNA-based nucleus). Since today's Achaeans present a chromatin-like chromosome (DNA + histones), it really makes a lot of sense that this group of microorganisms gave rise to the nucleus of the primitive eukaryote. Then, chronocyte engulfed heterotrophic aerobic bacteria (which later generated the DNA-based mitochondria). This theory is symbolized as E = A + B + C or in words Eukarya = Archaea + Bacteria + Chronocyte and is called the ABC hypothesis. Thus, this cell was the early eukaryote, which then evolved to form the earliest fungi and animal cells (Fig. 1.1b). Later, this early eukaryote cell engulfed a cyanobacteria (which therefore generated the DNA-based chloroplast), creating the first plant cell. The rest of the story is better known; once the eukaryote cells were formed, they turned into today's fungi, animal, and plant cells, forming unicellular or multicellular organisms.

Now, while bacteria and archaea generated a DNA-based genome (mainly circular chromosomes), chronocyte was probably an RNA-based genome organism since it evolved from its common ancestor acquiring a different strategy: ameboid-like cell with active motile behavior due to its cytoskeleton. In addition, it was probably able to undergo cell division without the need of a DNA-based genome, using an RNA-based centriole (a fundamental RNA structure in present-day eukaryotes used to organize the cytoskeleton during cell division). Moreover, chronocyte was able to synthesize and secrete proteins without the need of a DNA-based genome since only RNA molecules were required (and are required in today's eukaryotic cells) for this process, including ribosomes (complex macromolecules formed by proteins and RNA molecules), transfer RNA or tRNA (used to transfer the specific amino acids into the growing polypeptide chain), and messenger RNA or mRNA (which is used as template to transduce its sequence information into a linear chain of linked amino acids, the protein). In brief, this suggests that chronocytes were perfectly adapted organisms having an RNA-based genome that after a series of endosymbiotic events acquired mitochondria, chloroplasts, and a nucleus by engulfing bacteria, archaea, and cyanobacteria, respectively. Interestingly, later on, most of the RNA-based information was progressively transformed into DNA-based information and kept it into the nucleus. The process of turning RNA into DNA is currently called retro-transcription, since most organisms today undergo transcription that means the opposite, to produce an RNA molecule using a DNA molecule (chromosomes) as template. Probably, the retro-transcription was used early on to produce the common DNA-based genome bacteria-archaea ancestor from the original RNA-based genome common ancestor (see Fig. 1.1b). Considering that retro-transcription was probably an earlier process than transcription takes us into a next interesting issue: When did viruses appear? Probably very early on during evolution. For instance, early DNA-based genome organisms (bacteria-archaea ancestor) generated only DNA-based viruses, which are true in today's bacteria, an archaea. The same was true for early RNA-based genome organisms like chronocyte, since most current viruses that infect eukaryotic cells are RNA-type, which is consistent with the fact that viruses appeared as early obligated parasites, using the host's replicative capacity to proliferate and propagate. Thus, chronocyte was probably already infected by a type/types of early RNA-based virus and, as a consequence, created an anti-viral strategy, which is present in today's eukaryotic cells, named stress granules (an RNA/protein complex present in the cellular cytoplasm in charge of deactivating viral RNA) [6]. Finally, later on, when eukaryotic cells acquired a nucleus and the chromosome became DNA-based material, RNA-based virus used the "antique" retro-transcription activity to turn its RNA-based genome into a DNA molecule after infection (this is the reason why these viruses are called retroviruses). This DNA version of the viral genome was able to insert into the host chromosome and therefore to direct the viral replication from the host's chromosome, using its entire transcription (RNA synthesis) and translation (protein synthesis) machineries, shutting off the host cell metabolism, producing new viral particles, and taking control of the infection process.

Regardless of the model used to describe the origin of eukaryotic cells, the truth is that around 2 BYA, the primitive oceans started to be inhabited by an increasing number of all eukaryote cell types, many of them stayed as unicellular organisms until today, but some others engaged in a new challenge and underwent a series of transformations to become a multicellular organism.

Multicellular Life: From Colonies to Organisms

In a primitive world that was overpopulated with microorganisms, endosymbiosis provided the opportunity for those microorganisms to combine to each other in order to become more complex types of organisms with more efficient metabolism and greater genetic diversity—and therefore higher survival chances. With the exception of the original eukaryotic cells, the important thing for these cells was to evolve amidst a primitive ocean environment plagued with microorganisms while competing for space and nutritional sources. The question now is how these special organisms survived in that highly competitive primitive world. One interesting theory offers a partial answer to this question by suggesting that some eukaryotic cells stuck together after duplication by two basic ways, thanks to extracellular molecules that glued them to each other, and by forming a type of communication tubes (or holes) between cells to facilitate the intercellular transport of small molecules [7, 8]. This was possible because they were able to secrete proteins through their endoplasmic reticulum systems which were deposited at the cellular membrane or at the exterior of the cell. These secreted proteins kept the cells together by binding them directly such as proteins anchored at the external surface of cellular membrane (interacting between them promoting cells to stick together, such as cadherins[6]) or proteins inserted into the cellular membrane, forming hollow tube-like structures that interconnect cell cytoplasm as a way of intercellular communication. In this way, cells divide into two cells, and instead of dispersing into the environment, they maintained attached forming small clusters of colonies. Moreover, because the cells among the cluster/colony where interconnected across their cytoplasm, it facilitated their communication by exchanging nutrients, gases, or metabolic products—including toxins or signals.

In addition, other types of secreted proteins called extracellular matrix (ECM) proteins, most probably early forms of collagen-like[7] molecules, helped to keep the cells together by functioning as an intermediate molecular structure. Thus, by having a hydrogel-like property, ECM provided additional advantages including keeping a lot of water around the cell to prevent dehydration, to create a special mechanical protection as well as stiffness. As a consequence, the cells forming these early clusters or colonies could divide up their work: different cells evolved into special structures and functions to perform particular jobs. For example, the cells located on the organisms' surface, which faced the environment, likely served to protect other cells. The larger size of the colonies also meant that solutions had to be found to ensure nutrient intake, gas exchange, and excretion for all members of the colony. Increased coordination between cells of the colony allowed them to outcompete other colonies. Eventually, the colonies developed rudimentary structures, with the cells located in the inner part of the organisms specialized in structural and digestive functions, central nervous system functions that computed external stimuli, and other capacities.

These early multicellular organisms were small, but they were nevertheless considerably larger than their prokaryotic ancestors. With increased numbers of cell and extracellular matrices, these new organisms' bodies existed on a new scale (the centimeter scale) in an environment where bacteria and archaea

---

[6] Cadherin is a type of protein anchored at the external surface of the cellular membrane that interacts with other cadherins located in different cells (using Ca^{2+} as an intermediate) and promotes cell-cell interaction.

[7] Collagen is a type of protein fiber functioning as the main structural component in most connective tissues of the body. As a part of the extracellular matrix, it provides strength and elastic properties to tissues like the skin, cartilage, bone, tendon, heart, etc.

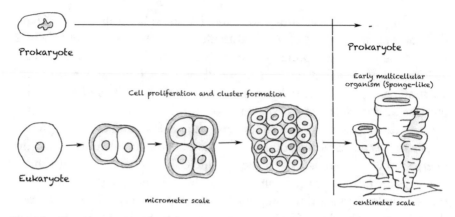

Fig. 1.2 *The new environment colonized by multicellular organisms.* Eukaryotic cells started to stick together after cell division, which led to the formation of primitive multicellular organisms. From their tiny origins, they grew in size until they became several times larger than prokaryotes and looked like sponges. Note that they went from micrometric scale (where 1 μm is equivalent to 1000 of a millimeter) to the centimeter (cm) scale (1 cm is 10000 times bigger than a μm). Thanks to its size, this early multicellular organism colonized and expanded without constriction in a new environment where unicellular prokaryotes were very tinny

looked like tiny, almost undetectable points (micrometer scale) (Fig. 1.2, *The new environment colonized by multicellular organisms*). Because they were much bigger, these organisms could interact with an entirely new and desolate environment where they were able to rapidly generate diversity at a completely new level. Fossil records indicate that these groups of early multicellular organisms that appeared around 800 million years ago (MYA) looked like sponges, with rudimentary symmetry.

Because physical space was not a constrain for these new multicellular organisms, their advent marked the beginning of an incredible diversification of species. In addition, although there was still much less O_2 in the environment than there is today, cyanobacterial metabolism increased levels of O_2 to allow the generation of multicellular organisms. Thus, these animals kept growing in size, generating more internal diversity, and acquiring multiple strategies to increase their survival fitness, like developing better feeding, camouflaging, or attacking strategies. In fact, all the unicellular organisms—including bacteria, archaea, and even unicellular eukaryotes—became their favorite meals, which was very beneficial for the multicellular organisms. Interestingly, some groups of bacteria and yeast adapted by invading their digestive track and establishing symbiotic[8] relation with them (microbiota).

---

[8] Symbiotic is the interaction between two types of organisms living in close physical association resulting in a beneficial relationship.

The Ediacaran Period

These groups of multicellular organisms, which were named Ediacaran biota after the Ediacaran paleontological period, appeared in the primitive oceans around 630–540 MYA [9]. The increment of the temperature in Earth due to the end of a glaciation period was considered the cause of the great diversity of shapes and sizes that appeared at that period, also named the Avalon Explosion. Their members displayed bilateral or radial symmetry, and they were planar (like leaves, ribbons, or tubes), likely because of the limitation posed by the low O_2 concentration in their tissues as well as their rudimentary feeding capacity (Fig. 1.4, *Organism diversity during Ediacaran period*). An initial hypothesis formulated by paleontologist working in this period suggested that most of the Ediacaran biota might not be considered as animals or plants as we know today; instead, it was believed that they were lichens, algae, protists, fungi, bacterial colonies, or some kind of intermediate between animals and plants. This conclusion was reached because they did not observe in their collection of fossil registries structures that resembled embryos or juvenile stages for members of all these groups. This, as expected, suggested that instead of developing into adult organisms after a process of fertilization, development, and maturation (like plants and animals do), the *Ediacaran* organism were basically colonies or clusters of individual cells forming complex structures, probably propagating by shedding pieces of their bodies or by dissociation, dispersion, and re-clustering. This, most probably, was the mechanism used to generate all the different types of organism found during this period. The question is how individual cells growing together, or getting associated by forming clusters, were capable of forming such diverse complex structures? Because if we look carefully, it seems that a very refined design is present in each one of them. The answer is simple: by an intrinsic principle that operates in nature called self-organization.[9] Self-organization in biology is a process that ends in the formation of patterns through interaction among the members of the system, such as cells [10]. The pattern is formed by the multiple direct interactions among the cells without external instruction or template. In other words, cells can form amorphous clusters after association but also by self-organizing and interacting themselves in different way would end in the formation of ordered and well-defined structures, like the examples observed in Ediacaran period. In this way,

---

[9] Self-organization is a process where a pattern at the global level of a system emerges solely from numerous interactions among the lower-level components of the system. Moreover, the rules specifying interactions among the system's components are executed using only local information, without reference of the global pattern [10].

many forms and shapes that appeared during this period might be the consequence of the direct incidence of the self-organization process.

Unexpectedly, an amazing discovery of new fossil specimens took place about 50 years ago (in 1971) at the Ediacara Hills of South Australia by an extraordinary man named Reginald Sprigg (an Australian zoologist with high interest in paleontology). Sprigg's new fossil records provided strong evidences suggesting that the first animals with sexual reproductive capacity appeared sometime during the Ediacaran period. This discovery completely changed the way of thinking about this period, and therefore, a new hypothesis emerged.

Self-Organization

It is a common process present in physical and biological systems, and some well-known examples could help to understand the significance of such phenomena, for instance, pattern of sand grains assembling into rippled dunes, pattern of cracks produced when mud dries and shrinks in a lake or pond (physics), fish schooling patterns, comb patterns in honey bee colonies, duck V-shape pattern in collective flying, termite mound-building patterns, myxobacteria feeding swarms, bacterial spatial patterns, butterfly's wings patterns, etc. (Biology). All these pattern examples, which most people know, are obtained by the process of individual interaction (exchange of information) among their members. In the case of the sand grains, the wind transfers the energy (which is dissipated through the system) necessary to make sand grains to order in such a way to form rippled structures. In the case of the fish schooling patterns, this dynamic process happens because the members of the group are capable of exchanging information very rapidly about the presence of a predator, which enables the entire school to execute a well-organized collective maneuver. Each member reacts in a way that indicates the rest how to react as well. There is no leader or predisposed behavior since all the pattern is dynamic and spontaneous, and since no prediction of the eventual presence of a predator occurs, the school responds uniquely as a group. The most relevant pattern formation to the potential self-organized process suggested during the Ediacaran period is probably related to microbial pattern formation. Myxobacteria are predators that feed on other microorganisms, and therefore, in certain circumstances, they self-organize in larger clusters or swarms that attack and surround the prey, like any known animal pack of predators will do with a prey. Additionally, bacteria, like the well-known *Bacillus subtilis*, and others, including *Paenibacillus dendritiformis* and *Paenibacillus vortex*, would form beautiful radial three patterns (very similar to some of the Ediacaran biota) when grown on agar plates under specific conditions, indicating that all the beautiful shapes observed during the Ediacaran period could very probably be due to a self-organizing process (see Fig. 1.3). Other examples follow basically the same principle, and the members of the group interact forming specific patterns without using blue prints or any type of external guidance. This conclusion is widely extended: from small sand grains to cell, tissue, or organism pattern formation.

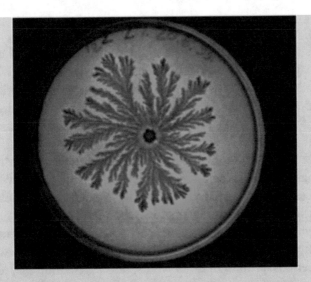

Fig. 1.3 Colonies of bacteria undergoing self-organization. The advanced collective and social behavior obtained by the well-known bacteria *Bacillus subtilis* clearly indicated that under special challenged culture conditions (consisting of low nutritional and dried environments), individuals within a colony might work together to survive. In this way, the colony is progressively developing where different bacteria is performing different tasks. This self-organized structure produces patterns that spontaneously form by the collective behavior of bacterial cells interacting. (Reprinted with permission from Ben-Jacob, E., Schochet, O., Tenenbaum, A. et al. Generic modelling of cooperative growth patterns in bacterial colonies. Nature 368, 46–49 (1994). https://doi.org/10.1038/368046a0)

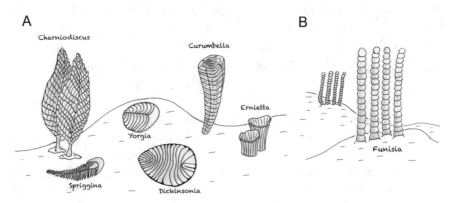

Fig. 1.4 Organism diversity during Ediacaran period. (**a**) The organisms indicated in the figure are some examples of the multiple that lived in this period. Note that these organisms looked very planar, similar to leaves, ribbons, or tubes, mainly due to the fact that molecular oxygen (O_2) concentrations were lower than they are today. (**b**) Another example of organism of this period is *Funisia*, which is considered the first evidence of an organism that had developed sexual reproduction. This assumption is based in the fact that all the colonies observed in fossil registries present individuals of the same size, which is common in organisms that produce offspring that are synchronically undergoing same developmental stages

Since this is widely known as the period when high diversity of multicellular organism shapes and sizes emerged, sex must have to be evolved. Why? Because sexual reproduction (which will be described in much more details in the following chapters) promotes the generation of high genetic diversity within the members of a population (belonging to the same species) and therefore increases the survival capacity of some more adapted individuals in case if a rapid change in the environment happens. In this way, the surviving members of the population would reproduce and generate offspring that would inherit the advantageous genetic characteristics. This is, basically, a simple definition of natural selection. In this way, gradually, as a consequence of natural selection, isolated members of the same species (i.e., exposed to different environmental changes) would diverge generating new forms of life (new species) in very short periods of time. In other words, self-organization played an important role in generating a great spectrum of organism forms and sizes but was not enough to provide the required genetic variability needed to have better chances to survive to the environmental changes, especially at the end of the Ediacaran period. Thus, it is postulated that those members of the Ediacaran biota that appeared during this period with an adapted body structure and, in addition, used sexual reproduction to generate offspring presented elevated chances to be the predecessor organisms of the next period, the Cambrian.

A fascinating story about Reginald Sprigg's discoveries was not realized until 2008 and indicated that his fossil records would be the key to identify the first multicellular organism that appeared on Earth with sexual reproduction. What happened was that a colleague and friend of Reginald Sprigg, Dr. Jim Gehling, announced the discovery of the origin of sex after analyzing in more details Sprigg's fossil records, as well as others collected by him at the same site [11]. The first organism discovered to have sexual reproduction was named *Funisia*, which was a worm-like tubular organism found abundantly at the Ediacara Hills. But how did they arrive to this conclusion by just looking into very old petrified fossils? Well, this is how they brought to light the information hidden in these samples. First, all the *Funisia* specimens collected were carefully measured and classified. Second, they realized that every time they found clusters of *Funisia* specimens, they were always of very similar size and growth stage (Fig. 1.4b, *Organism diversity during Ediacaran period*). Thus, this evidence made them reached to an astonishing conclusion: this would be expected from a primitive organism that after sexual reproduction, it produces many individuals that form new colonies and develop at the same time. Therefore, their growth stages are somehow synchronized. Now, if instead an organism reproduces asexually by budding (shedding pieces of itself), a wider range of sizes would be expected among juveniles. This process of mutual dropping sperms and eggs into the water occurs currently with corals, jelly-fishes, anemones, and even fishes, which indicates that this early way of sexual reproduction still remains in certain today's life-forms.

The Cambrian Explosion

By the end of the Ediacaran period (around 540 MYA), when all of this teeming life was still in the oceans, the O_2 level had risen due to the increase of the cyanobacteria population, and most of the Ediacaran creatures disappeared due probably to O_2-caused toxicity. Although it is believed that some Ediacaran members survived and contributed to give rise to more complex animal body architectures, feeding strategies, and survival adaptations, the fact is that a completely new set of animal forms appeared at that time. Regardless of how the new biota emerged, it clearly competed with the much simplest and rudimentary Ediacaran biota. In addition, it is also very likely that this new sophisticated biota was also responsible of causing their extinction by predating them. In other words, during the Cambrian period, the organisms generated more specialized tissues and organs for better protection, attack, movement, and detection (including sensorial organs), as well as better circulatory, digestive, and respiratory systems. At this time, a group of organisms started to be able to perceive light, shapes, and shadows, and the first primitive eyes appeared, providing a new, powerful survival strategy. Moreover, it is believed, but not proof, that other sensory structures including olfactory and tactile might have appeared at this time. This period, called the Cambrian Explosion (540–480 MYA) [12], saw the emergence of well-known organisms like ammonites and trilobites, which were the predecessors of all animal phyla that exist today, including mollusks, arthropods, and annelids (Fig. 1.5, *Animal diversity during Cambrian Explosion*). In 1920, a striking finding was reported in China (named Qingjiang biota) where a large number of specimens were collected including many organisms similar to jellyfish, anemones, worms, sponges, algae, and even arthropods. Curiously, in some cases, the internal body structures were so well-preserved that it was possible to identify special tissue and organ structures such as muscles, gills, mouths, guts, and eyes. This discovery certainly indicates that these fossil records were the direct predecessors of today's living fauna. Therefore, more recent periods witnessed the appearance of the modern animals, including insects, fishes, reptiles, amphibians, birds, and mammals of innumerable sizes and shapes.

In summary, the generation of the eukaryotic cell marked a major evolutionary step that enabled the production of the multicellular organisms that

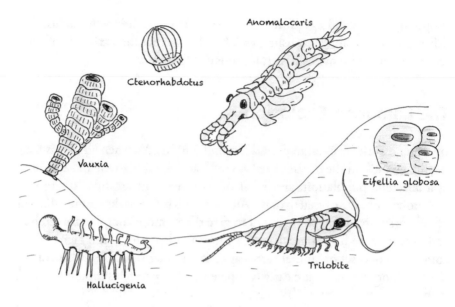

Fig. 1.5 Animal diversity during Cambrian Explosion. The animals indicated in the figure are some examples of the multiple that lived in this period. Note that these animals displayed more advanced body shapes and motility, as compared to the animals of the Ediacaran Period. Indeed, a much higher concentration of molecular oxygen (O_2) was present during the Cambrian Explosion, similar to today's levels, which helped to the development of more complex organisms

covered the planet (including all of its aquatic and terrestrial environments) with an incredible diversity of life, independent of the fact that the planet was already fully populated by bacteria and archaea. Hence, it is interesting to consider that multicellular organisms evolved exclusively from eukaryotic cells, since bacteria or archaea never contributed to forming multicellular organisms; instead, they eventually formed colonies or communities of their own, as Ediacaran unicellular organism did, but never evolved to develop into the next complexity stage: multicellular organisms reproducing sexually to give rise to the next generation. Paradoxically, the rise of eukaryotic multicellular organisms created an opportunity for unicellular organisms (mainly bacteria and fungi) to colonize new niches—and they did it in a very efficient way, as we know just by thinking about the trillions of bacteria and fungi living in our guts and all over our skin.

In conclusion, in this chapter, it can be stated that:

- Life started on Earth in the form of a primitive common cellular ancestor about 3.7 billion years ago (BYA).
- From that primitive common cellular ancestor, three main groups of cell types were generated, bacteria, archaea, and proto-eukaryotes around 2.5 BYA.

- Bacteria and archaea continued its own group generating thousands of species that currently live in any possible environment.
- The proto-eukaryote cell was probably an RNA-based genome organism, with ameboid characteristics including elastic body cell, having a cytoskeleton to generate movement and engulfing capacity and producing and secreting proteins using an endoplasmic reticulum-like, named chronocyte.
- First eukaryotes emerged by the endosymbiont process of chronocyte engulfing a heterotrophic bacteria (turning later into a mitochondrion) and an archaea (turning later into a nucleus).
- These first eukaryotes were the ancestors of the modern fungi and animal groups.
- Later, the first eukaryotic cell underwent a second endosymbiont process of engulfing cyanobacteria (turning later into a chloroplast). These eukaryotic cells were the ancestors of all modern vegetal groups.
- Eukaryote cells engaged in forming multicellular organisms (colonies) that formed complex structures, around 630–540 MYA, named the Ediacaran biota.
- Fossil records indicate that sexual reproduction of multicellular organisms started during the Ediacaran period, which caused acceleration in the formation of many species in a short period of geological time.
- By the end of Ediacaran period, most of its representative organisms were extinguished, most probably due to the atmospheric O_2 level increase or by the predatory behavior of a more sophisticated organism types that appeared at that time, giving rise to a new period named the Cambrian Explosion (540–480 MYA).

References

1. Mereschkowsky K (1905) The nature and origins of chromatophores in the plant kingdom. Biol Centralbl 25:593–604
2. Mereschkowsky K (1910) The theory of two plasms as the basis of symbiogenesis, a new study of the origins of organisms. Biol Centralbl 30:353–367
3. Margulis L (1970) Origin of eukaryotic cells. Yale University Press, New Haven
4. Hartman H (1984) The origin of eukaryotic cell. Specu Sci Technol 7(2):77–81
5. Hartman H, Fedorov A (2002) The origin of eukaryotic cell: a genomic investigation. Proc Natl Acad Sci USA 99(3):1420–1425
6. Zhang Q, Sharma NR, Zheng Z-M, Chen M (2019) Viral regulation of RNA granules in infected cells. Virol Sin 34:175–191
7. Grosberg RK, Strathmann RR (2007) Multicellularity: a minor major transition? Annu Rev Ecol Evol Syst 38:621–654

8. Niklasa KJ, Newman SA (2013) The origins of multicellular organisms. Evol Dev 15(1):41–52
9. Shen B, Lin D, Xiao S, Kowalewski M (2008) The Avalon explosion: evolution of Ediacara Morphospace. Science 319(5859):81–84
10. Camazine S, Deneubourg J-L, Franks NR, Sneyd J, Theraulaz G, Bonabeu E (2001) Self-organization in biological systems. Princeton University Press, Princeton
11. Droser M, Gehling J (2008) Synchronous aggregate growth in an abundant new Ediacaran tubular organism. Science 319(5870):1660–1662
12. Runnegar B (1982) The Cambrian explosion: animals or fossils? J Geol Soc Austr 29:395–411

2

Multicellular Organism Propagation

Summary The essential principle of life perpetuation through uninterrupted life cycles of individual organisms is described in detail. During this process, life flows by acquiring many forms such as the gametes formed by individuals, gamete fusion (egg fertilization), embryogenesis, growth and development, adulthood, and sexual maturation, and again, another cycle of life is initiated. Nevertheless, the key factor for this process to operate is based on the basic principle that individuals have a finite lifespan while life is propagated indefinitely.

> *Thus, from the war of nature, from famine and death, the most exalted object*
> *which we are capable of conceiving, namely, the production of the higher animals,*
> *directly follows. There is grandeur in this view of life, with its several powers,*
> *having been originally breathed into a few forms or into one; and that, whilst this*
> *planet has gone cycling on according to the fixed law of gravity, from so simple*
> *a beginning endless forms most beautiful and most wonderful have been,*
> *and are being, evolved.*
> —Charles Darwin, *The Origin of Species (1859)* [1]

Sexual Reproduction and Genetic Variability

Since the appearance of sexual reproduction, in principle, during the Ediacaran period, multicellular organisms gained the capacity of producing higher genetic variability in much shorter period of time which enabled them to rapidly evolve (in geological times) in multiple species. The fundamental principle of this mechanism is based on the development of a special cell type, the germline, which is basically able to generate gametes. Importantly, during the formation of gametes (gametogenesis), cells undergo a process of chromosome recombination, which produces multiple variants of each chromosome and, as a consequence, of new set of genetic variability. The fusion of gametes generates new organisms with unique genetic material and unique phenotype (genetic expression) which contributes to the heterogeneity among individual of a determined population.

Now, we know that cells do divide in two, whether they are prokaryotic, eukaryotic, or forming part of a unicellular or multicellular organism (Fig. 2.1, *Generation of cells by mitosis and gametes by meiosis*). Biologists describe this

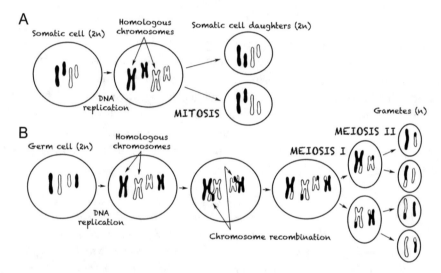

Fig. 2.1 Generation of cells by mitosis and gametes by meiosis. (**a**) Cell division of a somatic cell (2*n*) by mitosis of a diploid cell that has two homologous copies (one dark brown and one light brown) of each of the two different chromosomes (one type is large and another type is short). After DNA replication, each of these chromosomes is duplicated, and during mitosis, the duplicated pairs are separated again to create two new cells that are identical to the original one. (**b**) Meiosis, initiated in the germline cell (2*n*), follows the same DNA replication process as mitosis, but in addition, the replicated homologous pairs exchange pieces of DNA with one another in a process called homologous chromosome recombination. After this recombination, meiosis I divides the homologous chromosomes into two cells, and then meiosis II separates the sister chromatids again. The process reduces the number of chromosomes by half, producing four cells that each has one copy of each chromosome type, which represents the haploid number of chromosomes (*n*). These four cells are the gametes

process by saying that a mother cell gives rise to two new daughter cells. This particular cell division is called mitosis, a process in which the genetic material—the DNA, which takes the form of two pools of homologous chromosomes—is first replicated (duplicated) and then separated into two new cells. In this way, after mitosis, each daughter cell has the same amount of genetic material than the original mother cell. If we accept that the first organisms that appear on Earth were unicellular, then this way of proliferation (mitosis) to increase in individual number (expansion) was used for at least few thousand MY since the origin of life. In addition, the first multicellular eukaryotic organisms also used mitosis to help grow in size and bring shape to its early tissues and organs. Moreover, early eukaryotic organisms used what is called asexual reproduction in order to propagate. Curiously, this reproductive mechanism is still present in some of today's animal which includes few strategies such as binary fission (the process of splitting into two halves to form two new organisms, present in amoebas and in some sea anemones and coral polyps), budding (the outgrowth of a piece of an organism leading to the generation of a smaller new organism, present in corals and hydras), and fragmentation (the breaking of a part of an organism generating a new organism by regeneration, present in sea stars, earthworms, flatworms, sponges, etc.). Therefore, it is evident that the use of this expansion approach is very efficient in the way that each animal could have control of the time as well as the number of reproductive events in order to generate descendants, according to their needs. Now, there are no doubts about the speed and efficacy of the process, but from the evolution point of view, this reproductive strategy was not enough to produce the amazing amount of species diversity observed during Ediacaran period as well as the Cambric Explosion. Hence, something was missing, and independent of the animal size, a reproductive strategy for most eukaryotic organisms, called sexual reproduction, based on the production of single-cell gametes, emerged.

As briefly described in Chap. 1, the eukaryotic strategy of sexual reproduction tremendously increased the generation of genetic diversity in shorter periods of time (geological times). As a consequence, more heterogeneity among individuals (caused by the genetic diversity) of determined population (individual variants) would increase the chances that many of their members survive after a process called natural selection.[1] Why? Well, it is simple. Let's

---

[1] Natural selection is one of the basic mechanisms of evolution. In this process, genetic variability, differential reproduction, heredity, natural selection, and generational time determine whether an organism has a higher or lower chance of surviving and producing fertile offspring. The theory of evolution by natural selection was initially proposed by Charles Darwin in his book *On the Origin of Species*, published in 1859 [1].

imagine a large group of fishes, for instance, a school of sardines. Now, some members swim faster than others which will have more or less chances to survive a predator's attack. Thus, after subsequent attacks, it is expected that the slower fishes would have more chances to be eaten than the faster ones. In this way, natural selection operates by eliminating the members of the population less adapted to escape to the predator. Then, the survivor will have better chances to reproduce (differential reproduction) and as a consequence generate offspring. These offspring are the new generation, and they have inherited the genetic material from the survivor's group. Therefore, the new generation would have, in essence, individuals with swimming capacity more similar to the survivors (faster). Because this selective process is repeated all the time, the predator will also have to adapt to a constantly changing population of faster swimming sardines, and as a consequence, it would also promote the evolution into a better attack strategy. This process overtime contributes to the species evolution.

Although some simple organisms can reproduce asexually through fission, budding, or regeneration after fragmentation, most have access to reproduce sexually. Thus, it was clearly stated that the ability to reproduce by generating single-cell gametes and, as a consequence, being able to generate genetic variants certainly made a great impact during evolution. Moreover, the increasing competition among organisms as well as the highly changing environment of the Ediacaran period mainly provided selective pressure for sexual reproduction as an extraordinary survival strategy. But the question is: How sexual reproduction contributed during evolution to the high increment in genetic variability? The easy way to understand is by starting explaining what is a gamete and how are they generated. Now, gametes are the product of a process called meiosis, which was first described in 1883 by Belgium scientist Edouard Van Beneden [2]. It consists of a sequence of two unique cell divisions in which the genetic material—the DNA, which takes the form of two pools of homologous chromosomes—is first replicated and then separated, instead of mitosis, via two consecutive cell divisions, into four cells called gametes. Thus, each gamete will carry half of the initial genetic material. For instance, if we have a eukaryotic cell whose genetic material consists of two pairs of homologous chromosomes (two from the mother and the other two from the father), after the first meiotic replication, the cell will end up with four chromosomes—or two pairs of replicated homologous chromosomes—with each one containing sister chromatids (see Fig. 1.5b). Next, a cell division separates each of the replicated homologous chromosomes into two cells in a step called meiosis I or homologous segregation. A new cell division then splits each of the replicated homologous chromosomes into two new cells (this division is called meiosis II or sister chromatid segregation). As a result, each of the resulting four cells (gametes) contains a single copy of each chromosome type. Therefore, meiosis halves the initial number of chromosomes while maintaining a copy of each chromosome type. Thus, the number of different types of chromosomes is defined as the haploid number or n, and the total number of chromosomes of an organism that contains two copies of each chromosome type (or two homologous chromosomes) is defined as diploid number or $2n$. For instance, in the example of Fig. 1.5b, this cell has two different chromosomes, and therefore the haploid number (n) is equal to 2 and the diploid number ($2n$) to 4. Humans have 23 pairs of chromosomes, so their haploid number is 23, and their diploid number is 46, because their cells contain two full sets of chromosomes (one set from the mother and the other from the father). Moreover, gametes (n) are binary by nature, meaning that they are either male or female, and a masculine gamete will always fuse with a feminine gamete to generate a new diploid cell ($2n$), called a zygote. From this zygote, a new organism emerges (see Fig. 2.1, *Somatic cells and the germline of multicellular organisms*). Because we are used to think of human and animal individuals as multicellular organisms, it is difficult to see that part of a life cycle is as unicellular organism. These cells, the gametes, connect each of the multicellular life stages or what we use to call generations. Thus, life flows through each cell division (regardless of mitosis or meiosis), generating two cells. In some life stages, each of these cells is in fact an individual, including gametes or unicellular organisms (i.e., yeast, bacteria, amoeba, etc.), while in other life stages, these cells undergo several mitoses producing a multicellular organism. This multicellular organism will eventually have to die, but in order to maintain life flowing, it will generate gametes to start a new generation.

Natural Selection

Natural selection is one of the basic mechanisms of evolution, which is the change in the heritable characteristics (genetic material) of individuals of a specie over time (generations). In this process, genetic variability, differential reproduction, heredity, and generational time determine whether an organism has a higher or lower chance to survive by natural selection and, therefore, to produce fertile offspring. The theory of evolution by natural selection was initially proposed by Charles Darwin in his book *On the Origin of Species*, published in 1859 [1]. The central issue for species survival is based on four principles that work during evolution: (1) genetic variability, (2) inheritance, (3) natural selection, and (4) generational time. In other words, organisms must have mechanisms to generate genetic variability which is, in essence, a random process (i.e., mutations, genetic recombination, retroviral infection, etc., explained in detail later in this chapter). The generated genetic variability (in the form of some traits, like the example of the sardine swimming characteristics) is passed on from parents to offspring, which is the basis of heritage. Now, as much genetic variability a population has (a distinct phenotypic heterogenicity), more survival chances for certain individuals of the group would exist in a constant challenging environment (i.e., climate, food and water resources, predators, competition, space, etc.). Natural selection operates very simply: organisms that are better adapted to their environment will have more chances to survive and, as a consequence, to reproduce (differential reproduction) and generate offspring. Less adapted organisms would have less chances to survive and therefore to transfer their genetic material to offspring. Time is a key factor, since natural selection ensures that members of the population maintain constantly adapted to the challenging environment, but in order to work, it requires an adequate generational time, which for each specie is the time that an organism would need to reach sexual maturity since fertilization. This generational time could be hours or years, depending on each organism type, like unicellular bacteria or yeast, broccoli, or human beings. During the last 3.7 BYA, since the origin of the first primitive common ancestor cell, evolution has produced a huge biodiversity at any imaginable level of biological organization, including different types of organisms and species. Finally, species engage to different types of natural selection, and over time, they might undergo various patterns of evolution including convergent evolution, divergent evolution, and coevolution. Convergent evolution is the process that some species undergo that is not closely related and includes the generation of similar kinds of traits by independent mechanism. A good representative example is the wings of birds, insects, and bats, which are structures generated independently during evolution and are called analogous structures. Divergent evolution is the process in which the trait present in a common ancestor evolves into different related structures with different utilities. For instance, human arms, dolphin flippers, and horse front legs evolved from a common ancestor and are called homologous structures. Coevolution is a very interesting type of evolution where two species intimately related apply selective pressures on each other. Examples are normally observed among predator-prey pairs as well as species that use other species to facilitate reproductive fitness. The example represented in Fig. 2.2 illustrates the selective pressures among the predator (shark) and their prey (sardines). The shark will eventually catch the slower fishes of the fish school, as expected. On the other hand, the prey will also be selected by the shark feeding activity, and as a consequence, they will become faster swimmers over time. In this way, both species coevolve. Other interesting examples are present between pollinizing insects (bees) and their plant flowers they use to feed (nectar). Bees and flowers will coevolve in order to have better pollinizing strategies (highly furry legs in bees) and more attractive flowers for bees (more nectar, easy to obtain, etc.). In this symbiotic relationship, both species gain benefits; bees help flowers to reproduce, by spreading pollen from flower to flower; and flowers provide food (nectar) that the bees collect to feed the colony.

Fig. 2.2 A sardine school attacked by a shark. Some members of the school are faster swimmers than others, and therefore their chances to escape from the predator are higher. In this way, by constantly being challenged, the fish population is under natural selection. In this way, survivals would eventually procreate transferring their genetic material to offspring. In resume, evolution proceeds by changing the heritable characteristics (genetic material) of individuals of a group over time (generations). Artwork made by Claire Jarrosson Moral

Now, it is important to mention that during meiosis, right after the homologous chromosomes have replicated themselves, they engage in a very important process called homologous chromosome recombination [3]. During this process, the chromosomes exchange genetic material with one another to generate new chromosomes composed of mixed pieces of the original pairs (see Fig. 2.1b). For instance, in our case, humans received a maternal set of chromosomes that are the recombination result (the mix) of the previous chromosomes received by our mother parents (our grandparents). The same happens of course from our paternal set of chromosomes. Therefore, we, as sons and daughters, are the result of receiving the mixed chromosomes of our grandparents, and the "chromosome blenders" are our parents. In this way, each of the four gametes transfer new types of chromosomes—new genetic information—to the next generation. This process produces high genetic diversity and ensures that in each generation, the new zygote is formed not only through a random selection of maternal and paternal chromosomes (homologs) but also through the acquisition of new chromosomes built from recombined pieces of the original homologs.

In Chap. 1, the potential origin of retroviruses was discussed in relation to the origin of eukaryotic cells, in particular with the RNA-based genome common ancestor called chronocyte. These ancient retroviruses contributed with an important benefit to the genetic diversity of eukaryotic cells since after infection, its RNA-based genome was retrotranscribed into a DNA-based copy which subsequently was inserted into the host cell chromosome, generating random changes in the DNA sequence of chromosomes. These genomic alterations produced mutations on specific genes or regulatory sequences that ended in positive or negative changes. As we know, natural selection proceeded to maintain the beneficial changes and excluded the negative ones. Moreover, during evolution, these inserted retroviruses evolved into variants that were unable of completing the natural viral cycle of producing more viruses that ended killing the host cell, exiting and dispersing its progeny. Instead, they remained replicating internally generating more retroviruses-like DNA elements (called retro-transposable elements) that have the capacity to insert into other places of the same genome (chromosomes). This particular type of retroelements is denominated "endogenous retroviruses," which basically can propagate internally. It is, in a way, an internal self-infecting mechanism to produce random insertions. In humans, these elements are called HERVs (human endogenous retroviruses) [4]. In this way, HERVs progressively increase in number and therefore operates as an internal mechanism to generate genetic diversity. Currently, about 22 different families of HERVs have been identified in humans, which correspond to about 7% of the entire human genome. Interestingly, the detection of different retroelements (like HERVs) in our genome suggest that about 40 MYA, before the separation of Old World and New World monkeys, the primate germline experienced a sharp peak in retroviral integrations [5], implying that it might be one of the causes that generated the high divergence of primates, including the appearance (much later) of humans about 2 MYA. These suggest that HERV-like elements may have played an important role during evolution by increasing the genetic variability of eukaryotic organisms [6, 7].

In brief, organisms have several strategies to generate genetic variability to have more or less chances to survive during natural selection including mutation, recombination during meiosis, random chromosome distribution in the gametes, and fertilization chances among members of the same population.

Growth, Development, and Maintenance of Animal's Body

After egg fertilization (or ovule fertilization by sperm in mammals), embryogenesis takes place where our multicellular organism, which contains several different tissues and organs, must undergo an extraordinary process called development, growth, and maturation before it can be considered an adult. This issue was explained briefly in Chap. 1 when researchers elucidated the existence of the first organisms such as *Funisia*, which acquired sexual reproduction during the Ediacaran period. In this case, adult *Funisia* specimens were surrounded by many smaller exemplars of the same size, which indicated the presence of offspring generated by sexual reproduction (all eggs fertilized at the same time by swimming sperms), and at the same developmental stage, called juveniles. Thus, the process of development begins from one single cell (the new zygote). Through the proliferation process, or cell division, that first cell reproduces itself, and the organism's body increases in cell number and tissue mass. Progressively, the proliferating cells start creating new cell types through differentiation, undergoing embryogenesis until the organism becomes complex enough to be autonomous—a new individual, similar, yet different from its progenitors. This new organism is the new generation, also called the offspring. Then, the offspring reaches juvenile stage, which are smaller than their parents. In the case of *Funisia*, all the "siblings" grew synchronically, with similar body sizes. Juveniles require growth at first (production of more cells) and then maturation, which is the final stage of development, to a given structure-functional state. At maturity, the offspring also achieve the capacity to sexually reproduce in the same way their parents did before them, in order to continue life and pass it from generation to generation (Fig. 2.3, *Somatic cells and the germline of multicellular organisms*).

This brings us to the next issue: How do multicellular organisms decide which cells will contribute to the formation of gametes (germline cells) and which ones will contribute to the formation of the body or soma (somatic cells), including its tissues and organs? This decision is made very early in an organism's development when a small number of cells are selected to be part of the reproductive organs, including the gonads, which contain a special type of cells called the germline (this issue will be explained in detail in Chap. 3). Through meiosis, these germline cells generate gametes, which are essentially unicellular organisms that get together (during fertilization) to form another unicellular organism (the zygote). And through cell proliferation, this zygote generates a multicellular organism (the body)—which, once again, contains a

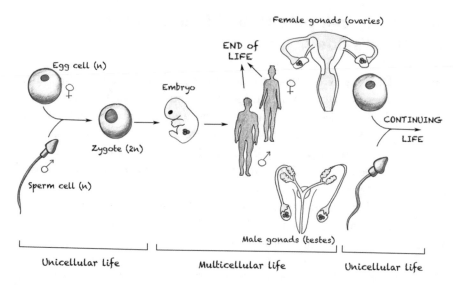

Fig. 2.3 Somatic cells and the germline of multicellular organisms. Life for multicellular organisms like humans is an alternating pattern: unicellular life (the gametes and zygote) is followed by multicellular life (the body or soma), which reaches maturity and generates more gametes that fuse to generate a new zygote, and so on, continuing the pattern of life. Gametes are haploid (*n*) but after they fuse to generate a diploid zygote (2*n*). The zygote develops into an embryo, where the germline is defined very early on. This embryo then generates the body that will be maintained until the end of its lifespan. The period from the early embryo until the body's adulthood is the multicellular life stage, which will in turn generate gametes from gonads to start a new unicellular stage (with a new zygote). This is why life is not created anew every time, but continued

structure (the gonads) dedicated to generate gametes that will fuse together to generate another zygote, which will develop into another multicellular organism, and so on, passing life on to the next generation. Thus, life is always continued; it is not created anew with every birth. Here is a simple way of visualizing it: imagine the body as a carrier for the germline, which, depending on how good this particular body is at finding a reproductive partner, will have a higher or lower chance of contributing its DNA to the next generation. From the evolutionary point of view, our social life, intellect, knowledge, and education—all our activities, in the end—are important factors that influence our chances of meeting the right reproductive partner and having descendants (i.e., offspring). It is strange thinking in this way, but humans, unlike most of the rest of animals and plants, do depend on these factors to meet their partner to create a family. We, as humans, are very complex specie, and in a certain way, we have "created" new rules that definitely affect our adaptation and evolution to this new environment we belong, our society.

Now, it is interesting to consider that even today, our generations are always connected by passing through a unicellular stage. This means that still, in spite of our complex multicellular organization, our social lives, our communities, and the diversity of our species, unicellular life is essential for everyone, indicating that single-cell organisms still play a crucial role in evolution. Moreover, the development of multicellular organisms in the Eukarya group (vegetal, animal, and fungi) probably occurred because selective pressure favored organisms that grew in size and colonize new environments over unicellular organisms, because large-body eukaryotes did not have to compete with other microorganisms (bacteria and archaea). Nevertheless, it was crucial for these organisms to maintain the unicellular life stage (gametes) in order to achieve sexual reproduction, which increased their genetic variability and survival fitness. Cells are the final process that ended in the origin of life, and regardless if we are talking about bacteria, archaea, or eukaryotes, unicellular or multicellular organizations, ancient or contemporary, they will represent the most important unit of life forever. Without any intention in oversimplifying the complex processes that fertilized egg have to go through in order to form a mature animal (plant or fungi), it is clear that cells are an amazing building block of life.

Now, let's talk for one moment about the body (soma) of the multicellular organism. As we commented in Chap. 1, if we only consider the number of cells needed to form this body, a human being of about 70 kg, for instance, will have about 37 trillion cells (37,000,000,000,000 or 3.7×10^{13} cells). This means that if we start from one single cell (zygote), it will need to go through many cell cycles (cell divisions of mitosis) to generate a human being. Because cells split into two new cells with each division, the number of total cells ($N = 3.7 \times 10^{13}$) will be equivalent to $N = 2^x$, where x is the number of generations required to reach 3.7×10^{13} cells (or N). By applying basic mathematics, we can calculate the value of x, which (in this case) is equal to approximately 45. This means that 45 generations[2] are necessary to produce that number of cells. In our simple model, though, the number of cells grows exponentially, which means that the total number of cell divisions (N_{cd}) is actually equal to the total number of cells produced (n) minus one ($N_{cd} = n-1$).[3] Thus, 3.7×10^{13}

---

[2] A generation, in this context, is defined as the process where each cell produces two cells after cell division. In this case, two cells are produced in the first generation, four cells are produced in the second generation, eight cells are produced in the third generation, and so on, following the equation $N = 2^x$, where n is the total number of cells produced and x is the number of generations.

[3] The number of cell divisions (N_{cd}) is almost the same as the number of cells (N) because in the first generation after one cell division, two cells are created ($N_{cd} = 2-1 = 1$). In the second generation, two new cell divisions are needed to produce four cells, which means that in order to produce four cells, the system has undergone three cell divisions (two new ones and the first one). Therefore $N_{cd} = 4-1 = 3$, and so on.

cell divisions are needed to form a human being. Even more, if instead of a human being we consider a blue whale that can extend as much as 100 feet (approximately 24–30 m) in length and weight about 180,000 kg (180 tons), the number of cells (if we assume that human cells are similar in size as whale cells) would be about 95 quadrillion cells (95,000,000,000,000,000 or 9.5×10^{16} cells). Thus, as we calculated for the case of human beings, 9.5×10^{16} cell divisions are needed to form an adult blue whale, which is an amazing number of cells considering that all derive from a single one, the zygote.

But these numbers of cells are not enough to build a human being or a blue whale because, in addition, we must also to consider the number of cells required to maintain a human body for 80 years of life (which is an arbitrary life expectancy given here as an example) or a blue whale body. In all organisms, tissues regenerate over time (this is also called tissue physiological regeneration or tissue homeostasis), and some regenerate faster than others (different cell types have different lifespan); therefore, the body is constantly producing new cells to replenish those that have naturally died. In addition, we know that once animals reach adulthood, their body mass remains nearly constant to the end of their lifespan (though old people tend to lose body mass). Thus, back to humans, to estimate the total number of cells produced during a person's lifespan, we first need to know what proportion of the body's cells make up each tissue, and we need to know the average time it takes for each cell to be replaced (i.e., the particular lifespan of each cell type). For instance, red blood cells constitute 84% of the total cells in a person's body, and they have a lifespan of about 120 days, which means that—on average—the number of red blood cells generated by the body triples every year. Therefore, the approximate total number of red blood cells produced by a 70 kg human being in 1 year will be $3.7 \times 10^{13} \times 0.84 \times 3 = 9.3 \times 10^{13}$. Over the course of 80 years, this number will be approximately 7.4×10^{15}, with an equivalent number of cell divisions. If we then factor in 100% of our prototypical human's cellular makeup by adding cells such as platelet cells (which make up 6% of the body's cells and have a lifespan of 10 days, so the body will generate approximately $3.7 \times 10^{13} \times 0.06 \times 2920 = 6.5 \times 10^{15}$ of them over its lifespan), intestinal and gastric epithelial cells (1% of the total number, with a lifespan of 5 days, generating 2.2×10^{15} cells), endothelial cells (2% of the total number, with a lifespan of 2 years, generating 3.1×10^{13} cells), skin epidermal cells (1% of the total number, with a lifespan of 2 weeks, generating 7.1×10^{14} cells), and so on, we can estimate that the total number of cells produced during the lifespan of a human body is roughly 2.5×10^{16} cells (with an equivalent number of cell divisions) (see Fig. 2.4, *Lifespan of different tissue cells in humans*).

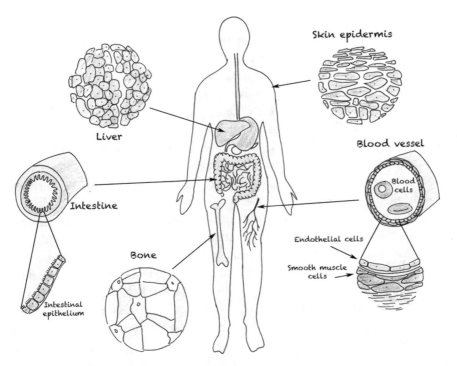

Fig. 2.4 Lifespan of different tissue cells in humans. Different adult tissues display faster or slower rates of physiological regeneration (i.e., tissue homeostatic regeneration). Intestinal epithelium takes only 5 days to regenerate, while platelet cells and the skin epidermis also regenerate quickly, every 10 and 15 days, respectively. Other tissues regenerate more slowly—red blood cells have a lifespan of 120 days; liver tissue lives for around 1 year; blood vessel endothelial cells have a lifespan of 2 years; and bone likely displays the longest regeneration time of all, with a lifespan of 10 years

Moreover, liver and pancreas tissues take around a year to regenerate, while fat, bone, and intestinal (non-epidermal) cells take 8–16 years. Thus, these latter tissues' contribution to the number of cells generated over a human's lifespan will be much smaller compared to other highly regenerative tissues, as described above. In any case, if cells did not regenerate, this astronomic number (2.5×10^{16}) would be equal to the number of cells in approximately 380 human bodies; in other words, it might be said that our bodies generate the equivalent of an entirely new human body every 2.5 months. Of course, this can be misleading, since we actually have some tissues that regenerate at a high rate (e.g., platelets, digestive tube epithelia, etc.) and others that regenerate at a medium (e.g., red blood cells, the skin's epidermis, etc.) or slow rate (e.g., the nervous system, bone, cartilage, etc.), as noted above.

A similar process occurs in the gonads, where the production of gametes through meiosis generates a quantity of unicellular cells that—depending on the species and its reproductive strategy—can also be astronomic. For instance, in humans, it is known that women produce around one to two million immature primary oocytes (egg cells) right after they are born, through a process named oogenesis. But by the time they reach puberty, this number drops to about 250,000 (2.5×10^5). These oocytes complete the first stage of meiosis right after puberty. Meiosis II only begins after ovulation, if fertilization (fusion with a sperm) occurs. Interestingly, in women, only one gamete (oocyte) survives after meiosis, while the other three degenerate and become the so-called polar bodies. Regardless of the fact that only one oocyte (instead of four) is produced in each meiotic cycle, women produce more than enough gametes, because about one oocyte emerges each month (during ovulation) until they reach menopause. In other words, around 500 oocytes are produced during a woman's fertile lifespan. Moreover, two concepts are interesting about oocytes: (1) it's a very long-lived cell since in humans, for instance, a 45-year-old women oocyte, it is truly 45 years old itself, and (2) it's a cell that spends its whole existence slowly transiting through meiosis. The situation is not the same for males, who constantly generate gametes (sperm) from puberty until the end of their fertile life. It is calculated that an average man can generate roughly 525 billion (5.25×10^{11}) sperm cells over his lifetime, which is a huge number of cells. Now, regardless of this difference in the number of oocytes and sperm produced by women and men, respectively, it is obvious that the gonads require special, controlled conditions during gamete production. This is a very important concept, because we are talking about the cells that have the essential job of carrying genetic information from one generation into another. In other words, the chromosomes that are transmitted through the germline must be of a very high quality, and they must have very few—almost no—changes (mutations). The genetic replication process and the cell division that occurs during meiosis operate under strict control to ensure that the products (the gametes) are, as much as it is possible, perfect candidates to generate a new and normal organism.

Although we will not discuss it in this chapter, it is important to mention that during regular cell division (mitosis), many changes in the genetic material including mutations, deletions of pieces of DNA, or translocations of pieces of chromosome among chromosomes may occur. This indicates that the cells that make up the body (with finite lifespans) are normal and functional, with good control of the fidelity of their DNA replication, but also that this replication is not as strictly controlled as gamete formation. In other words, the soma or body of the organism will have to eventually become less competitive, in order to give the younger members of the population the chance to procreate and produce offspring. Sooner or later, due to age, tissue and organ failure, or illness, the organism will die. On the contrary, the preservation of the germline, from its beginnings in the early blastula through to the formation of the gonads and the generation of gametes, requires that the body system protects its particular cellular lineage. In particular, to protect the germline, the genetic material is the key, since will be better done if the body system presents better survival fitness (including being a better hunter, having a better reproductive capacity, being faster and stronger, etc.). Thus, this is one the main factors for the species' survival during the natural selection process. In summary, the multicellular organism's body serves as a carrier that protects the small group of reproductive cells (the germline in the gonads) and ensures that the gonads are capable of generating gametes that will effectively find their counterpart gametes (from the other sex) to obtain fusion and generate a new unicellular organism: the zygote.

In conclusion, in this chapter, it can be stated that:

- During the Ediacaran period (630–540 MYA), multicellular organisms acquired the capacity to reproduce sexually which boosted the genetic variability among members of a population.
- Multicellular organisms therefore are composed of somatic cells (forming the body) and germline cells (in the gonads), dedicated to produce multiple single-cell gametes (oocytes and spermatozoids).
- Single-cell gametes are generated by a process called meiosis where a germline cell undergoes DNA replication and two consecutive cell divisions ending in four cell gametes carrying half of the initial genetic material present in the germ line cell.
- Importantly, during meiosis, homologous chromosomes undergo recombination ending in a high exchange of DNA that causes generation of new chromosomes (mixtures of paternal and maternal DNA). This is basically the mechanism that generates high genetic variability among organisms that reproduce sexually.
- During evolution, genetic variability, differential reproduction, heredity, and generational time determine whether an organism has a higher or lower chance to survive by natural selection and, therefore, to produce fertile offspring.
- Multicellular organisms produce an astronomic number of cells in order to develop a body (somatic cells) as well as to maintain it during the lifespan (physiological regeneration).

- Sooner or later, due to age, tissue and organ failure, or illness, the organism will die.
- On the contrary, the preservation of the germline requires that the body system protects its particular cellular lineage over life.
- Gonads of multicellular organisms will also produce large number of gametes along their fertile life to ensure procreation, which is generation of fertile offspring.
- Gonads are capable of generating gametes that will effectively find their counterpart gametes that after fusion generate a new unicellular organism: the zygote.

References

1. Darwin C (1859) The origin of species
2. Van Beneden E (1883) Recherches sur la maturation de l'oeuf et la fecondation *Ascaris megalocephala*. Arch BIoi 4:265–640
3. Lederberg J, Tatum EL (1946) Gene recombination in *Escherichia coli*. Nature 158:558
4. Herrera RJ, Lowery RK, Alfonso A et al (2006) Ancient retroviral insertions among human populations. J Hum Genet 51:353–362
5. Lander ES, Linton LM, Birren B, Nusbaum C, Zody MC, Baldwin J, Devon K, Dewar K, Doyle M, FitzHugh W et al (2001) International human genome sequencing consortium: initial sequencing and analysis of the human genome. Nature 409:860–921
6. McDonald J (1995) Transposable elements: possible catalysts of organismic evolution. Trends Ecol Evol 10:23–26
7. Brosius J (1999) RNAs from all categories generate retrosequences that may be exapted as novel genes or regulatory elements. Gene 238:115–134

3

Grow Fast and Well or Die

Summary Right after fertilization, most organisms' early growth and development in general need to happen very quickly. Why? Because most organisms are autonomous life-forms from the very beginning of their existence, like invertebrates, fishes, reptiles, amphibians, and birds. In this chapter, the main strategies for multicellular organisms to produce enough offspring that would contribute to the next generation and therefore to become fertile and procreate will be described. Most remarkable, despite of the evident diversity that exists among animal species (insects, mollusks, crustaceans, birds, reptiles, and mammals), the fundamental principles that apply during the reproductive process are very similar.

> *The succession of individuals, connected by reproduction and belonging to a species, makes it possible for the specific form itself to last for ages. In the end, however, the species is temporary; it has no "eternal life". After existing for a certain period, it either dies or is converted by modification into other forms.*
> —Ernst Haeckel, *The Wonders of Life: A Popular Study of Biological Philosophy (1904)*

Egg Fertilization and Growth

In the previous two chapters, we have introduced the idea of how unicellular organisms become multicellular during evolution and how these organisms started to reproduce sexually to pass life on to the next generation through an intermediary unicellular organism (gametes). This process occurs in the same way whether the next generation is made up of one single cell (yeast, amoebas, paramecium, etc.), many cells (as in insects or small invertebrates such as sponges, snails, worms, etc.), or a gigantic number of cells (as in reptiles, amphibians, birds, or mammals).

Most animals produce fertilized eggs at a rapid rate, and these eggs have to develop by themselves, with little protection other than being hidden, being many, or having an eggshell. This makes eggs and youngsters—in general— very vulnerable. Eggs are a very nutritious food, and they do not offer much resistance, so they are a good meal for hungry predators. Even so, egg deposition behavior in relation to care is different, for instance, between modern birds and fishes. Birds exhibit predominantly biparental care (90%), while in fishes, when it happens, it is mainly carried out by males (50%), followed by females (30%), and to a lesser extent also biparental (20%). In addition, many species of birds, fishes, or reptiles would protect their egg-containing nests very well showing an aggressive behavior against the intruder. Thus, it is an indication of the influence that "egg protection" has on increasing the survival chances, and vertebrates have developed strategies to protect their offspring. Moreover, this is not only true for vertebrates since it is well known what happens if one become very close to a bee, hornet, or wasp nest (hive) or, even worst, if you have the bad idea of touching it—the same for anthills, spider's webs, etc.

Now, regardless of the specific case, early animal development needs to operate at a very high velocity. Why? Well, because embryos need to produce a lot of cells in a very short period of time to ensure the generation of an autonomous animal that will be capable of looking for food, a place to grow, and a chance to escape from predators, increasing its chances of survival and reproduction. In essence, because natural selection operates insensibly during these early stages, an equilibrium between survival and death is well established. Thus, the better adapted offspring that would escape from predators implies that the population would have more chances to have enough number of individuals that have survived per generation. As a consequence, as much new survival members can flow into the population higher are the probabilities to contribute to maintain the genetic variability.[1]

Back to development, the simple task of producing a lot of cells in a very short time is actually a very complex process for two main reasons: first, before cells divide, they need to replicate their sizable genetic code (in humans, the

---

[1] In reference to an important concept cited by famous Russian geneticist Theodosius Dobzansky: "In biology, nothing makes sense except when seen in the light of evolution."

total DNA length per cell is about 1.8 m) in a way that minimizes the chances of generating mistakes (which are called mutations, genetic alterations such as chromosomal translocations, single-base alterations, deletions, insertions, etc.), and second, during these early stages, the organism must define its germline—i.e., the cells in charge of generating the gonads and, later, the gametes. Since mutations at this stage can arise from DNA replication mistakes (as well as errors in DNA repair processes during replication or environmental factors such as the presence of reactive radicals, oxidative stress, etc.), this means that early animal embryos must have a very good strategy for a quick developmental process while maintaining strict control on the fidelity of their genetic material (in order to reduce the mutation rate of the germline cell lineage). Given the fact that the germline is the source of all heritable genetic defects, it has a central role in the transmission of genetic material from generation to generation. In other words, in our picture of "the never-ending story of life," these cells are the vehicle responsible for the successful delivery of genetic cargo through the generations. We can say without doubt that germline cells are probably the most important cells that an organism might have. Why? Because while the somatic cells that complete the job of carrying these "vehicle cells" are of course also extremely important, they are nothing, really—from an evolutionary point of view—if the vehicle cells or the DNA they deliver are in bad shape. But more than being important for the individual, in fact, germline cells are more important for the survival of the individuals forming a population. In this game of surviving, adapting, and evolving, the individual is only a part of the real game, and the real "action" takes place only if the individual (the carrier) is able to transform the "vehicle cells" into gametes while is still young, but most importantly, it should find the right partner to mate and obtain fertile offspring. In this way, the task for an individual is to ensure that life flows, and it happens by producing as much offspring as necessary to contribute in maintaining the number of individuals in the population. Of course, the amount of offspring generated by each couple of parents varies in function of the type of organism, but the fundamental process is essentially the same for all.

The notion that germline cells are of supreme importance is supported by the data researchers obtained by comparing the mutation rate (the chances of accumulating mutations in the DNA) of various somatic cell types to that of germline cells in the same organism. This data allows us to draw certain conclusions: in mammals, for instance, the mutation rate for somatic tissues (per cell division) is generally higher than the mutation rate for germline cells. In mice, somatic cells mutate at about 2–10 times more than germline cells [1] and in humans 4–25 times more [2, 3]. Interestingly, in 15-year-old human beings, the chances of accumulating DNA mutations in somatic cells are around 10–100 times higher than in germline cells. Other organisms, like the insect fruit fly *Drosophila melanogaster*, present a mutation rate in somatic cells that is about 80-fold higher than in germline cells. Conceptually, it makes sense that somatic cells mutate more than germline ones, since in the end, the soma or body needs to play an important role, to ensure the transferring of the germ cells to the next generation through gametes that end forming offspring, but at the same time, it is important that the adult animal, once start aging and losing capacities, has to be replaced by young ones, which normally happens when old animals die. Moreover, somatic cells that normally accumulate more mutations than germline cells would increase their chances to accumulate defective genes that might cause cells to become malfunctioning. If this is the case, by time, the body starts accumulating malfunctional cells that progressively promote tissue and organ failure that leads to death. In short, organism bodies become aged, progressively turning less competitive among the younger members of the population to the point that are no longer competitive. This sounds tragic and unfair, and it is indeed, but is the way life flows: by giving a defined lifespan to individuals (mortality) as an exchange for giving indefinite lifespan to the germline cells that, by sexual reproduction, ensures population maintenance overtime, genetic variability, and adaptability. Of course, organism populations do not maintain themselves eternally unchanged through generations, but they evolve over time, changing, adapting, and generating more and more diversity—eventually, splitting into new forms and producing a vast spectrum of species. Life, therefore, continues through generations, through individuals that change shapes, numbers, and survival strategies, but it maintains over time, immortal.

The Life Cycle of the African Frog

Among vertebrates, one of the most spectacular developmental processes belongs to the African frog, *Xenopus laevis*. This amphibian[2] deposits its eggs in freshwater environments where they get externally fertilized by males. In only 36 hours (as long as the water temperature is between 23 and 25 °C), the eggs become small tadpoles that start eating and growing until they turn (in a week or two) into small froglets. During these phases, the organisms' body

---

[2]Amphibians are a group of animals that spend about the same time in water than in land and lay their eggs preferentially in freshwater environments which are fecundated outside of the body. Members of this group are frogs, toads, salamanders, and newts. It i0s a class of animal that presents gill respiration under water during an early stage of their development (larvae) and then replaced by aerial lung respiration in juvenile froglets and adults.

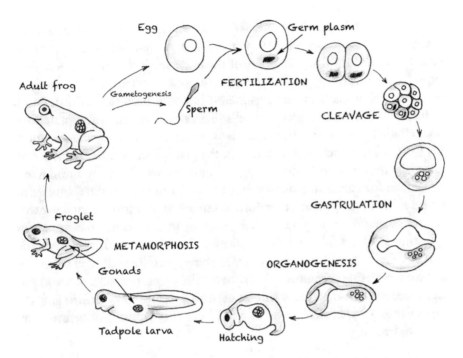

Fig. 3.1 The life cycle of the African frog. During the life cycle of the frog *Xenopus laevis*, eggs are externally fecundated by male sperm, forming the zygote that initiates the cleavage stage (a particular type of cell division). The egg already contains maternal germplasm that will be secreted through the development until hatching with a small group of cells (germline) that will determine the formation of the gonads in the small tadpole. This process takes around 36 hours. The tadpole then grows and undergoes a process called metamorphosis, turning into a small froglet. About 1 year later, it reaches its mature adult form. Adult frogs generate gametes (eggs by females and sperm by males) that will usher the life cycle into the next generation

structures change dramatically; froglets look completely different from tadpoles. This remarkable transformation is called metamorphosis and is observed in frogs (Fig. 3.1, *The life cycle of the African frog*) and in other organisms, like insects.

The froglets reach adulthood about a year later—but let's focus on the first 36 hours of the frog's lives, in which a little tadpole (of about 8×10^6 cells) is formed from one single fertilized cell (egg). Each one of these cells has about 3.3 rolled up meters of DNA that will need to be replicated every 25 minutes, at least for the first six to seven cell divisions. It is true that this gigantic amount of DNA is distributed across 72 chromosomes,[3] but still, each chromosome is very large (several centimeters long). As described above, the most amazing fact about this initial phase of the frogs' life is that the DNA replication must be practically perfect, because the germline (the cells in charge of producing gonads and therefore gametes) has only been defined just before the first cell division in a group of cells that carry a special cytoplasmic component called germplasm. In amphibians, as well as in fishes, insects, arachnids (spiders), and nematodes (worms), the germplasm is inherited from the maternal oocytes (non-fertilized eggs). These animals have an autonomous mechanism for regulating the presence of the germplasm in the future germline cells. Thus, germplasm, therefore, is crucial to "ensuring" that a very low number of mutations or genetic changes occur in these cells. In other words, it guarantees a high-fidelity DNA replication, much higher than in any of the somatic cells. Why? Well, as we discussed above, germline cells will be responsible for generating the gonads that contain the germ stem cells that will produce gametes. These gametes will be used to produce new offspring, and if these offspring were to carry a lot of mutations, they would bear defective and enviable descendants.

From the time the fertilized eggs become small tadpoles until they turn into small froglets (in a week or two), a huge cell proliferation activity is taking place. To achieve this fast, these animals need warmer temperatures to undergo active cell divisions and, as a consequence, to grow up fast. This would normally happen during spring season, when daylight hours help to warm up aquatic environments. In addition, predators (mainly fishes and reptiles) will also become active and hungry, which would be a problem for our little frogs. This is why, if early frog embryos do not grow fast and well, they are likely to be eaten by predators or die by not being able to survive by themselves. We can say definitively that the process of establishing the germline is crucial in guaranteeing life's continuation from one generation to the next. Nevertheless, it is important to understand that both germline and the animal body are associated to achieve good survival fitness. A reproductive strategy in frogs is to produce many eggs that are going to be externally fertilized by the male. This will generate a lot of swimming larvae that, most probably, will be eaten, but the fortunate ones, which undergo metamorphosis (see Fig. 3.2) and

---

[3] African frogs (*Xenopus laevis*) are tetraploid, which means they have 4 copies of each of the 18 different chromosomes as compared to humans, that are diploids and have 2 copies of each of the 23 different chromosomes.

Fig. 3.2 Transformations during the metamorphosis of the lime butterfly (*Papilio demoleus*). From left to right: larva or caterpillar walking onto the branch, early stage of pupa or chrysalid hanging (green), late stage of chrysalid hanging (colored), adult butterfly immediately after coming out of the cocoon, adult butterfly standing onto the branch. Artwork made by Claire Jarrosson Moral

reached froglet stage, would have better chances to survive. Of course, this is not the end of the story; still they are target of other predators, bigger fishes, reptiles (normally aquatic snakes), or birds. Even the adult ones are still in danger of being eaten by much bigger reptiles, birds, or even mammals. Thus, the life of a frog is not an easy one, and this is why it needs to reach adulthood (reproductive stage) during the next year, to be able to reproduce and start another life cycle. As we can see, these animals, as well as other amphibians, are subjected to constant natural selection, where the individuals with better camouflage strategies (able of blending with their environment, what is called biomimetic capacity) and better hunter techniques (mainly to catch small fish and insects) will increase their mating chances and therefore reproduce.

Metamorphosis

The word metamorphosis derives from the Greek word "μεταμόρφωσις" which means "transformation" or "change in shape." It is a biological process that occurs in many animals including some amphibians, fishes, insects, mollusks, and crustaceans but not in reptiles, birds, or mammals. The event normally occurs after hatching, which is when animals get out of their eggshell. The process is defined as an abrupt change in the animal's body shape by active cell proliferation, migration, and differentiation. The most spectacular example of metamorphosis known by everyone happens when a butterfly emerges from a chrysalid which is, without discussion, one of the most beautiful processes that can be observed in biology. The metamorphosis of moths and butterflies includes four stages: egg, larva, pupa, and adult. When the larva (also called caterpillar) hatches from its egg, it starts to eat and grow at such rate that in some species it can increase about 100-fold from its original size. They have tiny eyes, stumpy legs, and very short antennae. When the caterpillar stops eating, it turns into a pupa or chrysalid. In this stage, the pupa is covered by a cocoon (which can be of silk in most species of moths) and can stay there for weeks, months, or even years. In the outside, it seems that nothing is going on, but the truth is that major changes are actually happening. Cells are actively proliferating and differentiating, generating tissues that undergo abrupt morphogenesis forming the head, thorax, and abdomen as well as the legs, eyes, and wings of the future adult, the butterfly. The adults, as we all know, have long legs and antennae and composed eyes, and they can fly by using their large and, in most cases, beautiful color-patterned wings. Adult butterflies cannot grow, and their main job is to mate and reproduce, laying eggs for the next generation. Most adults can live from some weeks to few months. Monarch butterflies can fly between 50 and 100 miles per day (80–160 km/day), and for those that travel from Eastern United States to Mexico, the journey can last few months.

The Life Cycle of the Fruit Fly

The same principles apply to other animals, like insects. For instance, the most studied insect is the aforementioned fruit fly (*Drosophila melanogaster*) which also develops very fast. In only 12 days, mature adults arise from fertilized eggs. Fruit flies are only a few millimeters long, but they lay sizable eggs of about half a millimeter long. Once the eggs are fertilized, embryogenesis proceeds for 1 day, yielding a small feeding larva that grows for 7 days and then turns into a pupa (Fig. 3.3, *The life cycle of the fruit fly*). Four days later, a fly emerges from the pupa by metamorphosis. Like frogs, adult flies look completely different from pupas. In the case of flies, metamorphosis is as spectacular process as described for the case of butterflies but, unless you are a specialized biologist studying flies, is evident that observing a small dark fly to emerge is not the same as looking at the appearance of a beautiful colored butterfly.

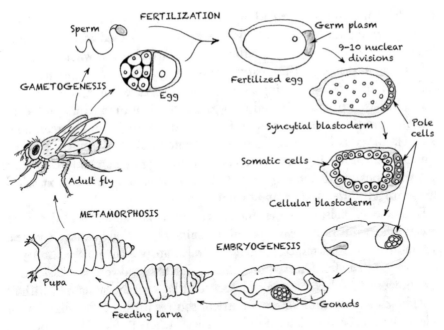

Fig. 3.3 The life cycle of the fruit fly. Adult female flies produce eggs, which are externally fecundated by male sperm, forming fertilized eggs containing maternally derived germplasm at one end. These eggs undergo between 9 and 10 nuclear divisions to form a syncytial blastoderm, which also contains pole cells that are defined by the presence of the germplasm. The syncytial blastoderm turns into a cellular blastoderm with the formation of a cellular membrane around the nuclei, on the periphery of the structure. Gastrulation promotes the formation of the organism's digestive tube (and main body) and the internal localization of the pole cells that will later define the formation of the gonads. By the end of embryogenesis, a small feeding larva hatches and grows, developing into a pupa, which turns into an adult fly through metamorphosis. Adult flies generate gametes (eggs in females and sperm in males) that pass life on to the next generation

During the first 3 hours of development, the egg's nucleus (which contains the fly's genetic material, arranged in eight chromosomes with about 8.4 cm of DNA per cell) divides itself 9–10 times until it has generated about 600 nuclei that share the same cytoplasm, called a syncytial blastoderm. This mechanism is quite unique among animals. The nuclei end localized near the periphery of the blastoderm, and the cell membranes develop around each nucleus, entrapping them into newly generated cells, producing what is called the cellular blastoderm (see Fig. 3.3). Those nuclei that enter into the germplasm will later give rise to the gonads and are called pole cells. During this period, the genetic material takes about 10–20 minutes to replicate; again, this organism must be able to generate perfect copies of chromosomes (its DNA) to ensure it will produce high-quality gametes. Then, the cellular blastoderm undergoes gastrulation which is a process where cells proliferate actively forming the first embryonic layers of tissues named ectoderm, mesoderm, and endoderm, which are the precursor tissues from where all tissues and organs will emerge. These early tissues engage in an active migration and differentiation process, forming a simple organism, called larva, constituted of the main basic structures including a gastrointestinal tube (from endoderm origin), a primitive nervous system (from ectoderm origin) in the main body, as well as the gonads (derived from the germline cells). By the end of embryogenesis, a small feeding larva emerges from the egg by hatching. Larvae are very voracious and therefore grow very rapidly, developing into a pupa, which turns into a fly after undergoing metamorphosis.

For these animals, the same rule applies, grow fast and well or die. Since insects in general are at the bottom of the animal food chain (since they are in general small, herbivores, etc.), their survival strategy is again based on laying a lot of eggs that after fertilization must reach adulthood very fast, faster than amphibians and fishes. Normally, insects that lay eggs in the nature, without providing any protection or extra nutrient sources further than what is available in the egg, produce large numbers of eggs, where the survival chances are purely based on the few ones that survive by chance. In addition, not only is impressive the amount of eggs that most of female insect can lay (between 200 to several thousand eggs) but also the incredible number of known species of insects on Earth, which is approximately one million and represents only 20% of the total world's species. The same gigantic number of terrestrial arthropods exists, being estimated a total (known and unknown) of seven million species. It is evident that insects and arthropods represent the animal groups with most diversity capacity and suggest that in the event of big environmental changes still a huge likelihood of survive exist. Interestingly, insects also have survival strategies that do not depend strictly in laying thousands of eggs. Instead, an example of some insects that only ensures to give birth to one larva at the time is the Tsetse fly. The larva is extremely well developed that it turns rapidly into a pupa and is ready to develop into an adult fly, giving to the offspring the best chances to survive, like mammals do, as described below.

The Life Cycle of the Mouse

In mammals like mice and humans, the process of generating and maintaining the germline is essentially similar, but it has some particularities that are characteristic of the mammalian class. First, the germplasm—the component of the cytoplasm that dictates how the germ cells form—is induced by neighboring cells. In other words, it is not determined by hereditary maternal factors in the oocyte, but rather induced later in the mammal's development, during the late blastula stage. Second, in mammals, egg fertilization, the progression of blastula stage until implantation in the uterine cavity wall, and embryonic development all take place inside the mother, between the fallopian tube and the uterus. This is a great evolutionary advantage because mammals are the so-called "warm-blooded" animal, which is to say that their body temperature remains constant under all circumstances (normally between 36 and 37 °C). Thus, intrauterine development proceeds in a highly controlled environment, and females can feed, migrate, and even escape from predators during the process (gestation), which helps them to better protect their future descendants (Fig. 3.4, *The life cycle of the mouse*). Third, once mammal embryos

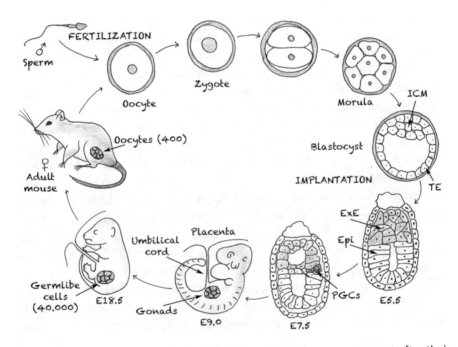

Fig. 3.4 The life cycle of the mouse. Adult female mice become pregnant after their oocytes (eggs) get fertilized by sperm during copulation. The fertilized egg turns into a zygote that travels through the fallopian tube, forming a blastocyst by cell division. The blastocyst, which contains the organism's trophectoderm (TE) and inner cell mass (ICM), is implanted in the epithelium of the female's uterus, where the TE cells contribute to the formation of the placenta (or extraembryonic tissues) and the ICM to the embryonic tissues. Once the blastocyst has been implanted, halfway through the fifth embryonic day (E5.5), the extraembryonic ectoderm—which is connected to the placenta—forms a cylinder-like structure with the cells derived from the ICM, which are called the epiblast (Epi). At embryonic day 7.5 (E7.5), around 20 progenitor germline cells (PGCs) emerge by induction between the extraembryonic ectoderm (ExE) and the Epi. Embryogenesis continues, and by embryonic day 9 (E9.0), the early gonads are formed. Female mice produce around 400 oocytes (gametes) during their lifespans

develop into mature fetuses, their mothers deliver them, and they begin their postnatal life, in most cases, surrounded and protected by their progenitors and, in gregarious species, in many cases by other members of the group. The reproductive strategy in this case is that embryos are in a privileged environment, protected from predators, and they do not need to find food by themselves for a while in order to grow and develop. They are mammals, so they feed mother milk from the mammary glands. This is true for mice, dogs, cats, monkeys, humans, elephants, dolphins, and whales. Some stay feeding from maternal milk longer than others, and since this food source is highly nutritious, it really helps the postnatal offspring to grow fast and strong, which is essential to be able to follow the group. This is important in migratory species, as it was important in early human groups of nomads, no so far ago in evolutionary times (50,000–100,000 years ago).

In mice, intrauterine life is very brief (about 19–21 days) compared to the intrauterine life of other mammals like humans (9 months) or whales (about 12 months). After copulation, fecundation occurs internally, where male spermatozoids, which are very motile, have to swim through the female's mouse uterus and the fallopian tubes to find the mature female oocyte (the egg) that has emerged from the ovary (ovulation). Normally, many spermatozoids reach the oocyte, but only one spermatozoid fuses with it forming the zygote. Afterward, the zygote travels in opposite direction than sperms, through the female mouse's fallopian tube finishing in the uterus chamber while undergoing a series of cell divisions (cleavage stage) that ends in the formation of a blastula stage, also called blastocyst. During this journey, which takes 2–3 days, the embryo experiences an amazing process which consists in an intimate contact with the maternal epithelium of the fallopian tube. Fallopian epithelium has many cilia (ciliated) at the cell surface facing the lumen of the tube, which move very actively helping the embryo to move through the tube in combination to the flow generated by tubal secretions. Additionally, the cilia movement touches the embryo providing a kind of "soft massage" that is transduced as a kind of mechanical communication among the maternal tissue and early embryo. Moreover, along the fallopian tube, the environment changes, which means that the embryo is exposed to different factors (like growth factors, cytokines, pH, etc.) that help in the developmental process from zygote to blastocyst. Apparently, this incredible process is very important since it establishes a communication between mother and its early embryo. Now, if we think for a moment, those embryos (blastocyst) that are obtained by in vitro fertilization (IVF) procedure would not actually have a maternal-embryo interaction like in the fallopian tube. This is because the formation of a zygote is artificially obtained by fusing a spermatozoid with an oocyte, and thus, the following development from zygote to blastocyst is carried in vitro on a dish. We don't know if this missing process might cause any difference in humans, but it would be, definitely, an interesting challenge in trying to recreate the "embryo journey" during IVF procedure using biomedical technology.

Now, the preimplanted blastocyst that had reached the uterus chamber presents two cell types, the trophectoderm (TE) cells and the inner cell mass (ICM) cells. Around embryonic day 4.5 (E4.5), the blastocyst gets implanted into the maternal uterine epithelia. After implantation, the TE cells will form a sheet of extraembryonic ectoderm (ExE), which is the embryonic cells that together with the maternal uterus epithelium will form the placenta (Fig. 3.4, *The life cycle of the mouse*). Moreover, halfway through the fifth embryonic day (E5.5), the extraembryonic ectoderm—which is connected to the placenta—forms a cylinder-like structure with the cells derived from the ICM, which are called the epiblast (Epi). The Epi has the potential to differentiate into all the somatic cells of the embryo and the germline cells. Then by embryonic day 7.25 (E7.5), signals from the ExE promote the differentiation of the most nearby Epi cells into germline cells, called progenitor germline cells (PGCs). From that point on, this group of cells is preserved to develop the gonads, ovaries in females and testicles in males, which will eventually produce oocytes and spermatozoids, respectively, once the mouse becomes reproductively mature.

Depending of the animal type, females can ovulate from one to several oocytes. Humans normally ovulate one oocyte per menstrual period (9–12 periods a year, for adult women) but could be eventually more. If women ovulate two oocytes and both get fertilized independently by two spermatozoids, then two zygotes will form, and as a consequence, two siblings will develop, which are called nonidentical twins. On the contrary, canine and feline females normally ovulate several oocytes (between four and eight in average per period) and therefore give birth to many puppies and kittens, respectively. Bu what are identical twins and how are they formed? It is simple. Briefly, after the formation of the zygote, a process of cell proliferation takes place in order to form the morula and then the blastocyst. Right after the first cell division, at two-cell embryo stage, if these cells for some reason split and get separated, then each cell will start its own process of cell proliferation forming two independent morulas and then blastulas. Each of these blastulas will then attach (implant) to the mother's uterine epithelium forming the placenta and the two embryos that later will form twins. Because these two siblings share the same genetic material, since both come from one single zygote, they are very similar or almost identical in appearance, and they are called identical twins.

Since the 1980s, scientists have learned how to cultivate cells isolated from the ICM of animal or human blastocyst. These cells can be disaggregated, loaded on cell culture flasks, and cultured in vitro relatively easily. After they attach to the plates, they proliferate very actively turning into what are called embryonic stem cells (ESCs). ESCs are normally cultured under special conditions and can be maintained indefinitely, which means that they have infinite lifespan or immortal. Scientist have been cultured ESCs for years and years without losing their proliferative capacity nor their unique capacity of producing many cell types after differentiation, which is called "pluripotency." But what exactly means that a cell is pluripotent? Stem cells are very primitive cells in general, and in particular ESCs are the most primitive ones. This is because scientists have performed experiments in the laboratory demonstrating that ESCs can be induced in vitro (by adding to the culture media diverse differentiation factors) to turn into many adult cell types including nerve cells (neurons), bone cells, cartilage cells, blood cells, pancreatic cells, etc. Since embryonic stem cells are originated from the inner cell mass (ICM) of the blastocyst, this implies that preimplanted embryos (the ICM) are immortal in essence. Later, after implantation into the maternal uterus epithelium, a process starts which progressively induces that most of the embryonal cells become differentiated (to form the body of the embryo), consequently, losing their infinite lifespan capacity. Only a small fraction of the original ICM cells will turn into germline cells (as explained above), which in essence are immortal. In humans, the intrauterine embryonic developmental process is basically the same, albeit with a longer time frame. As it was described in the mice model, the generation of the germline cells that will contribute to the formation of gonads and later gametes is produced after the blastocyst is implanted in the uterine epithelium. This is a new and important concept, since in fact immortal lifespan of germline cells seems to be associated with the immortal property of life itself.

Now, the concept of infinite lifespan of certain cell type, or what can be said "immortal" capacity, does not imply that the cell does not suffer mutations (genetic changes) over generations. In fact, as it was explained before, genetic changes are fundamental to generate diversity and therefore essential for evolution. In essence, all genetic material suffers mutations over time, different mutation rates, depending of the system, but there is not scape to that. Moreover, change itself is an important property associated with organisms' chromosomes, regardless if they are bacteria, fungi, plants, or animals. The

The Garden of Eden and the Immortal Life

Remarkable as it sounds, an incredible similitude exists between the immortal property of life itself (described in this book) and the well-known description of the Garden of Eden.[4] In this narrative, Adam and Eve were free to eat everything except the fruits from the tree of knowledge of good and evil. The garden was apparently a place of perpetual joy without pain, sickness, and, eventually, death. As we all know, Adam and Eve were seduced by the serpent to eat the forbidden fruit of knowledge, and therefore, they were expelled from the garden as a punishment to disobey God's commands. As a consequence, they have to subsist by themselves in our world, without protection, becoming susceptible to hunger, sickness, aging, and eventually death. Since they were no longer immortal, they have to procreate to generate children, to multiply forming a tribe, later a society. In other words, they lost their immortality, but something immortal was left in them (may be by an act of God's mercy), the germline cells. Besides all the unfortunate consequences that their act caused, life found their path, maintaining itself immortal through Adam and Eve's germline cells. Cells that after generating gametes and fusing gave rise to the first zygote, which grew to become a blastocyst, carrying immortal inner cell mass cells, that after implantation in Eve's uterus became a mortal offspring, susceptible to sickness, aging, and death, but carrying an immortal subset of cells, the germline. This is exactly what happened to their parents after being expelled from the Garden of Eden. This cycle was repeated itself countless times until present, and hopefully, it would be infinite, which reflects the wisdom encrypted in ancient narratives. Again, individuals are mortal but life is immortal.

point is not really about the change itself but instead about the mutation rate. If the mutation rate in the genetic material is high, then the organisms might suffer big changes in their genetic material (genes), and the products of these genes, proteins, could have extensive changes that might produce massive dysfunction and eventually death. If mutation rate is very low, then the organism (and the population associated) might have low genetic variability, and their chances to survive after significant environmental chances are, eventually, low. The mutation rate is precisely adjusted in each adapted organism and depends on many factors but mainly in the capacity to found dynamic equilibrium between change and adaptation to their environment. The main cells involved in transmitting these genetic chances through generation to generation are the germline cells since somatic cells do not transmit their genetic material to the next generation. This is why germline cells are very well protected in the

---

[4] Garden of Eden (Hebrew: גַּן־עֵדֶן), also described as the Terrestrial Paradise or the Garden of God in the *Book of Genesis* and the *Book of Ezekiel*. Similar description is present in the Sumerian story of the place of the immortals, where sickness and death were unknown. Moreover, it is also described in the Garden of Hesperides in Greek mythology similar in a way to the Jewish's Garden of Eden.

gonads and undergo very controlled cell division in order to generate the gametes that best adjust to the survival necessities of each species. Therefore, we can say that germline cells are the most important cells to ensure the future of the particular organism population, in terms of adaptation, survival, and evolution.

Interestingly, it has been studied—in a very precise way—the mutation rate of male and females, and it has been estimated that the mutation rate is higher in paternal DNA contributions than maternal counterparts by about 75–80%. Moreover, paternal age is an important parameter that can further increase the number of mutations that might be transmitted through the gametes to offspring [4, 5]. The reason is simple: the male germline undergoes a significantly higher number of cell divisions than the female germline, and as described above, DNA replication during cell division is one of the main sources of mutagenesis. Moreover, the majority of cell divisions in the male germline take place in the spermatogonial stem cells, which are the specific stem cells that produce sperm in the gonads during meiosis. Since men continuously produce sperm from puberty onward, this explains why the sperm quality changes over time (through the process of aging). As noted in Chap. 2, it is estimated that an average man can generate nearly 525 billion (5.25×10^{11}) sperm cells during his fertile life—a number that is about 1 billion times greater than the 500 egg cells generated by the average woman.

This clearly suggests that in humans, parental age is a crucial factor in ensuring the generation of normal offspring. Historically, emphasis on youthfulness has been highlighted in women because their fertility is limited by menopause, but more recently, it has become clear that the father's age is also a very important aspect. This makes perfect sense since humans, unlike other mammals or even other animals, do not normally procreate as soon as they reach sexual maturity (during adolescent age), which is usually not happening in our modern society. But this "natural timing" to procreate was fundamentally set during evolution by a simple biological reason: at this age, men and women present the capacity to generate gametes with a very low number of mutations (low mutation rate). Thus, shorting the generation time to 20–25 years would ensure that the progeny will have better survival fitness. This was the approximate generation time in humans during Neolithic times (12,000–6000 YA). But we know that our generation times today are much longer than the Neolithic, by many reasons including economic and social stability of the progenitors, need on finishing education, etc. Nevertheless, regardless of our modern culture conventions, we cannot escape from the biological rules without consequences. This, in our case, is transduced to the fact of producing offspring at much older ages as our own biology dictates, putting in high risk the fidelity of the genetic material of our next generation. This is why an important message for us to learn from our Neolithic ancestors is that we should perform—once it is decided to become parents (specially, after 30 years old)— a genetic test from the amniotic embryonic tissue. This test is fundamental in order to know potential chromosome abnormalities of the embryo, as well as point mutations, microdeletions, insertions, etc., which might help to know if the baby is likely to have certain medical condition. More importantly, in the latest years, genetic testing has advanced to the point that doctor's early diagnose of a potential risk can help to treat some illnesses and increase the quality of life of a person. Moreover, many illnesses would develop by the combination of carrying high-risk genes (special mutations in a certain genes) and environmental factors (the lifestyle factors such as eating, drinking or smoking behaviors, etc.). If the person knows his/her predisposition, he/she could decide in making lifestyle changes to avoid becoming sick before it is too late.

In conclusion, in this chapter, it can be stated that:

- In multicellular organisms, after the fertilization stage, growth and development need in general to happen very quickly to ensure high rate of survival during development.
- Early on during embryogenesis, the germline is defined to ensure the development of the gonads that will give rise to gametes.
- The mechanism of germline specification is different depending on the animal group. For instance, in the animal models described in this book, insects and amphibians differ significantly from mammals, like mice.
- Amphibians, fishes, insects, arachnids, and nematodes have cells carrying a special cytoplasmic component called germplasm from the beginning of embryogenesis. The germplasm is inherited from the maternal oocytes (non-fertilized eggs) and specifies the formation of germline cells.
- In mice, the germline is formed later in embryogenesis. Few days after embryo implantation in the maternal uterus, a group of about 20 cells emerges by induction signal secreted by cells from the embryo as well as from the extraembryonic tissue (embryonic cells that together with maternal cells form the placenta).
- Germline cell mutation rate (point mutations, microdeletions, etc.) is much lower than somatic cells.
- In humans, since the male germline undergoes a significantly higher number of cell divisions than the female germline, the chances to transmit a mutation to the next generation are three times higher.
- Parents older than 30 years should, at early pregnancy stages, perform a safe genetic test to the amniotic tissue of the fetus to know about potential issues in the genetic material of their baby.

References

1. Dycaico MJ et al (1994) The use of shuttle vectors for mutation analysis in transgenic mice and rats. Mutat Res 307:461–478
2. Lynch M (2010) Rate, molecular spectrum, and consequences of human mutation. Proc Natl Acad Sci USA 107(3):961–968
3. Lynch M (2010) Evolution of the mutation rate. Trends Genet 26(8):345–352
4. Kong A et al (2012) Rate of de novo mutations and the importance of father's age to disease risk. Nature 488:471–475
5. Rahbari RR et al (2016) Timing, rates and spectra of human germline mutation. Nat Genet 48:126–133

4

Why Do Animals Grow, Age, and Then Die?

Summary As simple as it sounds, the main purpose of the multicellular organism—that is, the body of an animal—is to ensure that life will continue by ensuring its transfer to the next generation. Nonetheless, from the biological point of view, life is long enough for most organisms to have a chance to grow, mature, learn, have experiences, and mate with the right reproductive partner. If this last thing occurs, the organisms can produce descendants to enhance chances of perpetuation. Once this is finished, each organism has accomplished their main purpose on life, and therefore, their remaining time will depend on how fast age, become dysfunctional, and die.

> *To an extent that has surprised us and the rest of the scientific community, telomeres do not simply carry out the commands issued by your genetic code. Your telomeres, it turns out, are listening to you. They absorb the instructions you give them. The way you live can, in effect, tell your telomeres to speed up the process of cellular aging. But it can also do the opposite.*
> —Elizabeth Blackburn, *The Telomere Effect: The New Science of Living Younger*

Tissue and Organ Homeostasis

We all acknowledge the arrival of a baby in our family or a friend's family as something normal and, somehow, expected. We also accept that children grow up—that they turn into teenagers and then into adults. Moreover, we accept aging as a natural process, and we accept the problems associated with getting old, such as reduced physical capacities and memory, certain tissue and organ failures, and so on. Nevertheless, we have a hard time accepting death probably because the concept of being transient carriers of life instead of immortals is commonly avoided, at least in occidental culture. But from the biological perspective, death is an important factor to ensure that the circle of life never ends.

In Chap. 2, we saw that multicellular organisms like humans have a tremendous capacity for regenerating tissues during their normal lifespan, but the question is *how* that regeneration happens. In addition, in Chap. 3, we were introduced to the concept of stem cells through our discussion of gametogenesis and their role in germline determination among vertebrates. But as we consider the somatic tissues and organs that protect these germline cells, a few things are important to consider now. Firstly, we should remember that each tissue requires a different cell division kinetics to maintain its cell mass over its lifespan. Secondly, each tissue must have its own unique strategy for replacing cells that die naturally while maintaining its tissue shape, size, and function. Not so long ago, it was considered intuitive to think that when a tissue cell died, it was replaced directly by the closest neighbor. The thought was that the neighboring cell entered into the cell division cycle, producing two daughters, one of which would replace the dead cell. But it turns out that this is not the complete picture. Although physiological regeneration follows this intuitive mechanism in some adult animal tissues (like the liver in mammals), in many others, the mechanism is not exactly like this. Instead, these latter tissues have specialized cells that live in specific niches and act as "selected cell generators." These specialized cells are called stem cells (SCs), and their niches—naturally—are called "SC niches." SCs are very special, because they are primitive versions of the particular cell type (or types) that make up a given tissue, and they remain primitive throughout the entire life of the organism. They are frequently referred to as "immature" cells, because they are capable of producing progeny that has the potential to turn (differentiate) into tissue-specific cell type.

For instance, if we are talking about the epidermis, which is the external part of the skin (see Fig. 2.2), the resident SCs in that tissue are called basal cells. They received this name because their niche is located at the bottom (i.e., base) of the epidermis, which is attached to a complex protein network called the basal membrane (BM) (Fig. 4.1, *Skin epidermal stem cell niche*). The BM is a very strong barrier that prevents epidermal cells from passing to the other side, the internal part of the skin, which is known as the dermis. On other words, BM separates the skin in two, the dermis, which is in contact to other internal tissues of the body and therefore is all blood-irrigated (or

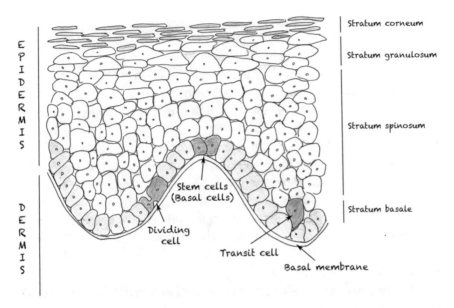

EPIDERMIS

DERMIS

Stratum corneum

Stratum granulosum

Stratum spinosum

Stem cells
(Basal cells)

Dividing
cell

Transit cell

Stratum basale

Basal membrane

Fig. 4.1 Skin epidermal stem cell niche. The skin's epidermal tissue regenerates continuously, using niche stem cells to produce new cells. These stem cells (basal cells) are located at the basal membrane (BM), and they form the stratum basale, which separates the epidermis from the dermis. During tissue homeostasis (physiological regeneration), basal cells divide asymmetrically to produce one basal cell and one progenitor cell. The latter continues dividing to form transit-amplifying cells that move through the BM until they become detached and migrate into the tissue (as indicated by the black arrow) to terminally differentiate into keratinocytes which constitute the cell layers including the stratum spinosum, stratum granulosum, and stratum corneum

vascularized) tissues inside the organism, and the epidermis, which is considered external and avascular (not blood-irrigated). Moreover, on the dermis side of the BM, the cells (called skin fibroblasts) are responsible for conditioning the stem cells' microenvironment by secreting niche factors (mainly proteins) that help the basal cells maintain their original properties, thereby ensuring that the tissue keeps their regenerative capacity intact. SCs have a peculiar cell division kinetics: after they divide, one daughter cell remains identical to the original SC, and the other becomes what is called a progenitor cell. This cell division kinetics is called asymmetric cell division, and it ensures that the original SC renews themselves after every division. The other daughter—the progenitor cell—then undergoes several more cell divisions to replace dead cells in the tissue with new, functional ones. When the progenitor cell undergoes cell proliferation, the new cells differentiate into the tissue's final, mature cell types. For instance, in the skin's epidermis, progenitor cell descendants will differentiate into keratinocytes, which is the main cell type of this tissue, and therefore the basal cells are considered to be "unipotent stem cells" (see Fig. 4.1). In this way, even though the SC always remains inside their niche, the progenitor cells generate functional adult cells as a mechanism of tissue regeneration. Normal human skin epidermis regenerates every 2–3 weeks, which indicates a high tissue turnover. Thus, a kinetic model for epidermal physiological regeneration indicates that if we consider 1 mm² of epidermal skin, approximately 1200–1400 cells are proliferating every day at the stratum basale [1]. This means that the same number of cells will become progenitor cells that therefore will migrate to the upper layers turning into terminally differentiated keratinocytes. These cells transit through the stratum spinosum to the stratum granulosum and finally to the stratum corneum. Thus, the same number of cells is exiting the skin every day to maintain the tissue homeostasis. In other words, the cells constituting the skin behave like a flowing system where the bottom produces cells that move to the surface (see the black arrow in Fig. 4.1).

This is one strategy by which adult animal tissues maintain their regenerative capacity while also maintaining tissue structure, tissue function, and cell mass. Importantly, this mechanism minimizes the number of cell divisions that the SCs undergo during an organism's lifespan, which is a crucial factor for sustaining the tissue's regenerative capacity over many years of life (in case of humans).

Other types of tissue—like blood, for instance—use the same strategy to produce all cell types that make them up. Although the blood stem cell niche (which is called the hematopoietic stem cell niche) is different from the skin epidermal one, its operating principles are very similar (Fig. 4.2, *The hematopoietic stem cell niche*). The development of the hematopoietic system is defined as hematopoiesis. In humans, hematopoiesis begins very early during development in the yolk sac, which is a membranous structure outside of the embryo. This primitive hematopoietic system involves a simple erythroid progenitor, with the function of producing red blood cells (erythrocytes) to facilitate tissue oxygenation while the embryo undergoes rapid growth. Now, these erythroid progenitors are transient since they do not have renewal capacity and therefore are not maintained for longer. Then, definitive hematopoiesis occurs later in development and takes place in the aorta-gonad-mesonephros region of the developing human embryo. It involves the formation of hematopoietic stem cells (HSCs), with multipotent capacity to generate all blood lineages of the adult organism. Then, at the sixth week of gestation, HSCs undergo temporal transition in the embryonic liver, staying there until the fifth month of gestation. At this time, HSCs colonize definitely the fetal bone marrow, which will be the final location throughout life.

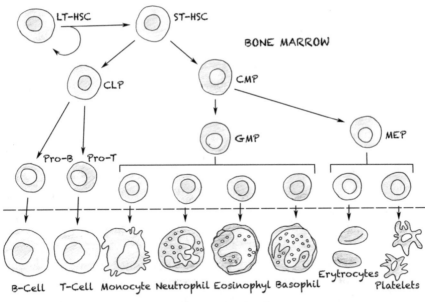

Fig. 4.2 The hematopoietic stem cell niche. The bone marrow contains a long-term hematopoietic stem cell (LT-HSC). Through asymmetric cell division, this cell generates another identical LT-HSC and a short-term hematopoietic stem cell (ST-HSP), which is the progenitor cell that will then generate two other progenitors through another asymmetric cell division: the common lymphoid progenitor (CLP) cell and the common myeloid progenitor (CMP) cell. The CLP generates prolymphocytes B (Pro-B) and pro-lymphocytes T (Pro-T) that will later differentiate into lymphocytes B and T (B-cell and T-cell), respectively. The CMP undergoes another asymmetric cell division to produce the megakaryocyte-erythroid progenitor (MEP) cell and the granulocyte-monocyte progenitor (GMP) cell. The MEP goes on to generate platelets and erythrocytes, while the GMP generates monocytes, neutrophils, eosinophils, and basophils. Monocytes will later form macrophages and dendritic cells (not shown). Finally, the GMP goes on to generate monocytes, neutrophils, eosinophils and basophils. All of these terminally differentiated, mature cells leave the bone marrow and enter into the peripheral bloodstream

During adult hematopoiesis, the HSC is able to generate both another identical (self-renewing) cell that remains in the niche and a progenitor cell that will continue dividing to generate a hierarchy of cells that will eventually differentiate into all cells present in the bloodstream (lymphocytes, monocytes, leucocytes, macrophages, red blood cells or erythrocytes, platelets, etc.). The HSC niche consists of a special microenvironment located at the bone marrow. There, HSC are in close proximity to vascular sinusoids that are expansions of interconnected arterioles and capillaries forming an interconnected network [2]. Sinusoids are particular structures that present high level of permeability because their walls are formed by a not so tight monolayer of endothelial cells. Thus, sinusoidal niches promote diverse functions of HSCs such as self-renewal, proliferation, and differentiation.

As mentioned in the last chapter, this niche is responsible for producing almost 90% of the total cells in the human body, which means it must be very active as well as long-lasting. It produces a gigantic number of cells following the same strategy as the skin's epidermis, thereby minimizing the number of times the HSC undergo cell division and maximizing cell division in the progeny cells.

Here, again, this mechanism guarantees that the organism is able to keep its original HSC intact while producing the functional blood cells it needs over the course of its lifespan. This mechanism is called physiological regeneration or tissue homeostasis, and it constantly occurs (at different rates depending on the tissue) all over the organism.

These last two cases exemplify the typical tissue homeostasis model, which is executed from the stem cell niche. This model is common to most of the body's tissues, but there are some exceptions, like the liver. This organ displays a remarkable mechanism whereby a particular population of mature hepatocytes (the cell type that makes up 70% of the liver) is able to divide by themselves. In the liver, these hepatocytes, rather than stem cells, maintain organ homeostasis (Fig. 4.3, *Liver physiological regeneration*). This requires a radically different strategy of tissue-organ maintenance, because hepatocytes have very high metabolic activity (elevated levels of oxidative stress, exposure to

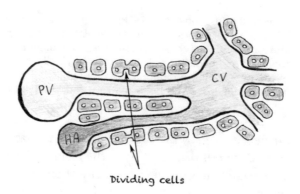

Dividing cells

Fig. 4.3 Physiological regeneration in the liver. During liver homeostasis (physiological regeneration), hepatocytes located anywhere along the portal vein (PV), hepatic artery (HA), or central vein (CV) can undergo cell proliferation to replace dead cells. The figure represents the minimal organ functional unit, in a very simplified way for didactic purpose

xenobiotics, metabolic derivatives, etc.). In the liver, where hepatocytes have a medium lifespan of 400 days and there is no stem cell niche, the organ must alternate among the hepatocyte population that is engaged in cell division in order to maintain the proliferation rate per cell at a minimum and achieve maximum efficiency. Although all mature hepatocytes are capable of engaging in proliferation, it seems that different subpopulations of hepatocytes display a particular aptitude for cell division [3]. Moreover, the liver is the organ that is best at regenerating after an injury (which we will discuss in the following chapter). It is intriguing to consider why this regenerative process also relies on hepatocytes rather than niche-protected stem cells.

Of course, if we were to continue going through the entire human organism, tissue by tissue, we would find similar stem cell strategies to those displayed by the skin and the blood, as well as more particular strategies like that of the liver, but we would quickly find that each case is different and unique.

It is curious to think that tissue stem cells are single cells operating alone among other cells (progenitor cells, niche cells, etc.). In fact, they are unicellular entities playing a strategic role in tissues' multicellular unity. This remarkable particularity of multicellular organisms—which allows them to gain access to regeneration—once again demonstrates that individual cells (in appropriate microenvironments) have evolved in such a way that they can have total control of the functioning and activity of a multicellular organism. We therefore can imagine a multicellular organism as having multiple "cell generators" dispersed throughout its tissues. These cell generators keep the organism functional by replacing cells as they need to be replaced. The same analogy can be extended to the germ cell line, where a germ SC niche keeps these cells protected within the gonads to ensure the generation of gametes during the fertile lifespan of the organism. Obviously, in the case of the liver, the model is different, and it seems that mature cells alternate with one another, taking turns participating in organ homeostasis and post-injury tissue regeneration.

Although the existence of tissue stem cells answers the question of how tissues and organs are able to regenerate so efficiently (and generate so many functional cells over an organism's lifespan), it does not answer the following questions: Why do we age, and how? In theory, the stem cell strategy described above should be able to maintain and regenerate an organism indefinitely, because every time a cell needs to be replaced, some specific stem cell (or hepatocyte) can produce that very cell as its progeny. But in practice, this turns out to be true for a while—for many years—but not forever. The reason is that cells, in general, suffer changes in their genetic material (mutations, deletions, and other genetic modifications) after each cell division, and these changes accumulate over the years, especially in those cells that proliferate most, such as progenitor cells, transient cells, and adult differentiated cells. This is a natural process, and the fact that those mutations occur in cells that will eventually die and get replaced tells us that adult somatic tissues can accumulate a relatively low number of mutated cells without much effect on their functioning. The problem starts after several years, when mutations and other genetic changes accumulate enough that they begin to affect stem cells. Once stem cells change, tissues' regenerative capacity can be affected—by a diminished rate of cell production, by the production of lower-quality cells, and by the accumulation of living cells that do not function well. Indeed, the same law of damage accumulation through prolonged proliferation applies to hepatocytes in the liver. Aging, in effect, is the direct consequence of the additive effect of all these cellular dysfunctions. In other words, tissues start showing evidence of age when their regenerative capacity starts to progressively decline while they accumulate nonfunctional cells (Fig. 4.4, *Tissues accumulate nonfunctional senescent cells over time*). Moreover, we know that adult cells that undergo more than a certain number of cell divisions are also susceptible to aging, because they can self-induce a program called senescence (which is the technical term for cell aging). The senescence that extensive cell division

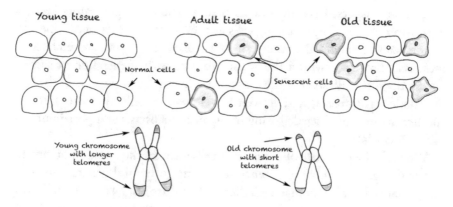

Fig. 4.4 Tissues accumulate nonfunctional senescent cells over time. In a hypothetical tissue, senescent cells accumulate over time. This can be observed by looking at a tissue at three different stages of its life (young, adult, and old). Normal cells are functional, with long telomeres and little DNA damage, while senescent cells have short telomeres and a high level of accumulated DNA damage. Essentially, an increase in these senescent cells will promote progressive loss of function in the tissue or organ

induces in an adult cell is called proliferative senescence, and it is normally activated after several generations (e.g., for the skin's dermal fibroblasts, senescence is activated in vitro after about 45 cell divisions or generations). One cause of proliferative senescence is related to the fact that the extremes (or tips) of each of the lineal chromosomes (called telomeres) get a little shorter with every mitotic cycle. In early cells (i.e., embryonic, fetal, or newborn cells), telomeres are at their maximum size, but after several cell divisions they become progressively shorter. When they become too short, cells have trouble replicating the tips of their genes, and the cell's senescence program is activated automatically. This indicates that cells have a molecular clock that has evolved to prevent the organism from accumulating adult cells that have divided too many times and, in this way, protect it from the risk that cells with so many mutations might produce aberrant cells and eventually tumor cells. Interestingly, the telomeres of some cells, such as early embryonic cells (from the zygote to the blastula stage), do not become shorter with each cell division, because they express an enzyme complex called telomerase, which re-lengthens the telomere to its original size after cell division [4]. For this reason, these cells do not suffer proliferative senescence or age in the same way and, as explained before, can be cultured in special conditions that mimic their natural microenvironment and would maintain their infinite proliferating capacity in vitro. Other groups of cells that maintain their telomeres intact in every cell division are the germline cells. It is evident why, since these cells can't undergo senescence and neither age, therefore a mechanism to maintain germ line cells "young" by activating telomerases is present at any stage of the organism's life. Moreover, some adult stem cells present more or less amount of telomerase activity, which indicates that their molecular clock works slowly compared to terminally differentiated tissue cells. This suggests that multicellular organism in general presents cells that age a different time at the body or soma (adult stem cells and differentiated tissue cells) and others that directly do not age at all, including germline cells. Thus, we can define in a more complex way an organism life by saying that the cell lineage in charge of producing gonads and gametes (germline) does not age and, as a consequence, life itself do not age but change during time by having different forms, adapting to changing challenging environments while perpetuating. Instead, individuals do accumulate progressively senescent cells that at certain point lead to tissue and organ failure and death.

Telomeres and Telomerases

Eukaryotic chromosomes are very long linear segments of DNA in intimate association with multimeric proteins named histones. In this way, the histones have two main functions: (1) to help packing the DNA in a tight structure called chromatin and (2) to mark on the DNA molecules what sections of the DNA sequence contain genes that might be expressed (euchromatin) as well as other segments that are never expressed (heterochromatin). Chromosomes have two defined heterochromatin zones: the centromere and the telomeres (see below). The centromere is an internal zone that is used during chromosome segregation in mitosis and meiosis, to attach the spindle machinery that will be used to separate each of the new recently replicated chromosomes (or sister chromatids) to each of the new daughter cells. Telomeres instead are zones at each end of the chromosome which provides protection, and the cell uses to count how many times the chromosome has replicated (which will be equal to the number of cell divisions). For instance, human telomeres in early embryonic cells present 2500–3000 times the same repetitive sequence, 5′-GTTAGG-3′, which is equal to 15,000–18,000 base pairs. After birth, telomere lengths have decreased to 11,000–14,000 base pairs and keep progressively decreasing during life. It is estimated that about 200 base pairs are lost in each DNA replication event because the enzymes involved in the process cannot replicate all the way to the end of the chromosomes. Therefore, after about 50 cell cycles, telomeres get too short (4000 base pairs or less), and a senescent program is activated causing little by little cellular dysfunction and cell death. This process does not occur simultaneously to all the cells of tissues and organs, since not all cells have undergone the same number of cell cycles. This, therefore, implies that tissues are formed by a heterogeneous population of cells with different "cellular ages," but in addition, it is also true that as organisms get older their tissues and organs accumulate more and more senescent cells. These cells, depending of the tissue, take more or less time to die and, as a consequence, are the main source of tissue or organ failure at old ages (see Fig. 4.4). Now, some cells can escape of becoming senescent after exceeding the normal limit of cell cycles. Some of these cells are the cells present in the preimplanted embryo forming the inner cell mass (ICM) of the blastocysts (from where the embryonic stem cells (ESCs) are isolated) and the germline cells. The way they bypass senescence is by expressing an enzyme complex called telomerase, which re-lengthens the telomere to its original size after cell division. Because the DNA replication machine cannot complete the double helix at the end of the eukaryotic linear chromosomes, the telomerase enzyme complex is in charge of maintaining the original DNA length. How it does so? The telomerase complex in humans is basically formed by two molecules, a telomerase reverse transcriptase (TERT) and a telomerase RNA (TR), which is a noncoding RNA. The telomerase thus uses the first RNA bases of the sequence 3′-CAAUCCCAAUC-5 to bind to the end of the non-extended DNA molecule (see Fig. 4.5). In other words, the telomerase uses the RNA as a template to extend the synthesis of DNA, and this is why it is called reverse transcriptase, because it synthesizes DNA from a sequence of RNA (this is reverse to the canonical transcription process, where the DNA is used as a template for the synthesis of RNA). Telomerase would repeat this process many times (about 30–35 times) in order to extend the DNA enough so then the DNA replication machine would be able to synthesize the missing fragment of DNA. In this way, the end of the linear chromosomes (telomeres) gets extended to their original size, and therefore, non-replicative senescence is activated, regardless of the number of cell divisions that the cell underwent.

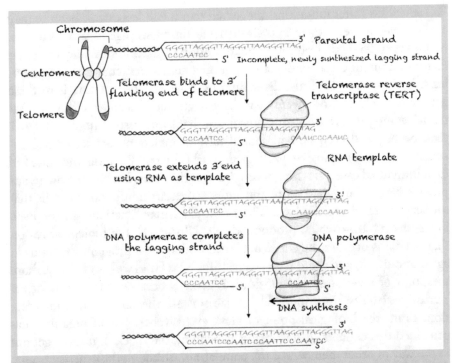

Fig. 4.5 Telomerase elongating a telomere. The telomerase enzyme complex (TERT and TR) binds to the parental DNA strand of a telomere using the complementary sequence present in the telomerase RNA (TR) template. The rest of the TR sequence is used as a template to synthesize DNA by the reverse transcriptase activity of the TERT enzyme. As a consequence, the DNA gets extended. In the figure only, one elongation cycle is shown for simplification, but actually many cycles normally happen (around 30–35, not shown), which produces DNA elongation of about 200–250 base pairs. This extra DNA length is enough for the DNA replication machine (represented as the DNA polymerase) to bind and to synthesize the complementary DNA chain and, as a consequence, to regain the original DNA length

In addition, cells can also become senescent before they reach the usual number of generations if they are exposed to damaging environmental factors (toxins, a high level of free radicals, UV light, elevated temperature, etc.), which produce accumulation of mutations more quickly. If this happens, in order to protect the organism, the cell will once again activate an internal program to induce senescence or, eventually, what is called programmed cell death or apoptosis. In essence, under these extreme circumstances, adult cells would start a senescence program early on, which continues throughout the entire lifespan of the organism. In other words, we slowly start aging as soon as we are born, but it is more evident as we become adult and, later, elder.

Tissue SCs suffer mutations, too, albeit to a much less degree because they are very well protected within their niches against aging agents such as toxins, high oxygen levels that promote the formation of free radicals, and any other factor that might cause modifications in their genetic material (reactive chemicals, UV light, etc.). Thus, depending on how exposed a tissue is to these genotoxic factors, those agents will have a greater or smaller effect on that tissue's aging process. In theory, this means that some tissues should age faster than others, and each human should have a predetermined set of circumstances that favor aging or prolonged youthfulness. Whatever the status of each individual case, one thing is certain: over the years, our tissues and organs progressively accumulate more and more senescent nonfunctional cells that sooner or later cause the fatal failure of a particular function—in the liver, lungs, heart, etc.—which produces death. Ultimately, the biological reason that we die is because the entire organism undergoes a program that ends in aging and death, and there is no escape from it, as far as we know today. From the point of view of evolution, death is crucial, because it is important for genetic variability since individual organisms are the units of natural selection. From the biological point of view, exclusively, once an organism has generated descendants (few or many, depending the species), it must get out of the way of other members of the population. Although the genetic changes that occur with each generation are almost undetectable, it is important for the species to have the chance to generate more variability over and over. This variability is obtained by constantly replacing the old players with new ones and by reducing the generation times to favor a decent degree of mutation rates, which is the basis of the strategy of survival called natural selection.

In my experience, though, biology always surprises us with new unexpected discoveries that make us reformulate our theories. This is also true for the main concept described in this book—"the never-ending story of life"—which refers mainly to the fact that life itself is immortal, even though each individual life is not. Unexpectedly, it turns out that some time ago, scientists discovered just such an exception: an organism that is technically immortal [5, 6]. The medusa jellyfish *Turritopsis* sp. can essentially live forever, though its form changes drastically over time (Fig. 4.6, *Regular life cycle and reverse cycle of a special jellyfish*). This organism does not live eternally as a mature jellyfish; instead, it displays a novel strategy: it leads a normal life until it reaches maturity, but when it is facing old age or sickness, it can revert to sexual immaturity through a process that involves a return to its early developmental stages. Eventually, it reverts all the way to its juvenile form and becomes a polyp again. In other words, instead of aging and dying once it has reproduced and propagated a new generation (like all the other multicellular

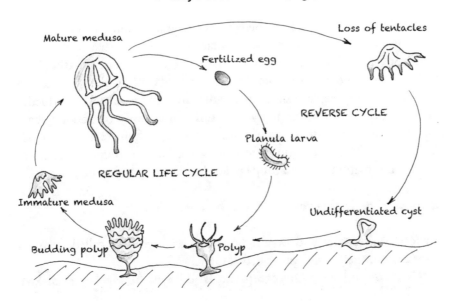

Fig. 4.6 Regular life cycle and reverse cycle of a special jellyfish. This particular jelly-fish, *Turritopsis* sp., presents two possible alternative life cycles: regular and reverse. The regular life cycle starts when adult medusas liberate eggs that have been externally fertilized by sperm. This fertilized egg develops into a swimming planula larva. At this stage, the larva adheres to the seafloor to produce a polyp, which generates other polyps forming a pile-up structure from which budding polyps gradually escape. This swimming structure is an immature medusa (called an ephyra) that eventually grows into its mature form. Alternatively, if adult medusas do not die and reach the brink of old age, they can opt to go through a reverse cycle, which consists in a drastic bodily transformation. By the end of this reverse cycle, the organism has once again become an undifferentiated cyst that sticks to the seafloor. From that simple structure, a new polyp emerges to reinitiate another regular life cycle

organisms we have discussed so far), the jellyfish's entire organism decides to deconstruct its whole body in an organized way. It loses its tentacles, and then its body shrinks until it becomes a small undifferentiated cyst that sinks to the ocean floor and forms a new polyp. Then, from that polyp stage, a new cell cycle restarts from the beginning.

This amazing organism actually proves the rule no organism can live forever in its adult and mature form. In most cases, apparent signs of aging mean that an organism's time is starting to get short and that the changes in its cells will eventually lead to death. In exceptional cases, like that of the special jellyfish, mature life is different, since the organism can choose to return to the earliest stage of its life. This is essentially a programmed form of individual survival that is distinct from the asexual propagation strategies of other previously studied organisms (i.e., amoebas' fission or the ability of earthworms, flatworms, and sponges to regenerate after fragmentation). In the end, this sort of immortality is reserved for very simple organisms that can begin life anew from a bunch of cells without sexual reproduction. The situation is certainly not the same for more complex organisms like fishes, birds, reptiles, and mammals, whose propagation occurs rigorously through sexual reproduction. For most organisms, immortality only exists in the way this book's central concept defines it: as life that continues through an infinite chain of individuals, flowing from generation through generation, through fecundation, growth, and maturation and propagation before aging and dying.

In conclusion, in this chapter, it can be stated that:

- Most adult tissues and organs present specialized cells called stem cells (SCs), and their niches are called "SC niches."
- SCs are primitive versions of a particular cell type (or types) that make up a given tissue, and they remain primitive throughout the entire life of the organism.
- SCs have asymmetric cell division kinetics: after they divide, one daughter cell remains identical to the original SC, and the other becomes what is called a progenitor cell.
- Asymmetric cell division kinetics ensures that the original SC renews itself after every division, while the other daughter—the progenitor cell—undergoes cell differentiation into a mature cell type.
- Skin epidermis and hematopoietic tissue follows the SC kinetics described above, while liver presents a different one.
- Mature hepatocytes are capable of engaging in proliferation; it seems that different subpopulations of hepatocytes display a particular aptitude for cell division during physiological regeneration.
- Adult cells that undergo more than a certain number of cell divisions are also susceptible to aging, because they can self-induce a program called proliferative senescence.
- Proliferative senescence is related to the fact that the extremes (or tips) of each of the lineal chromosomes (called telomeres) get a little shorter with every mitotic cycle.
- When telomeres become too short, cells have trouble replicating the tips of their genes, and the cell's senescence program is activated.
- Organisms accumulate senescent cells (nonfunctional cells) progressively during life until their tissues and organs stop functioning (tissue or organ failure) and die.

- Telomeres of some cells, such as early embryonic cells and germline cells, do not become shorter with each cell division, because they express an enzyme complex called telomerase, which re-lengthens the telomere to its original size after cell division.
- The medusa jellyfish *Turritopsis* sp. can essentially live forever, since once an adult is facing old age or sickness it can return to the earliest stage of its life to restart a new life cycle.

References

1. Weinstein GD et al (1984) Cell proliferation in normal epidermis. J Investig Dermatol 82:623–628
2. Boulais PE, Frenette PS (2015) Making sense of hematopoietic stem cell niches. Blood 125(17):2621–2629
3. Chen F et al (2019) Broad distribution of hepatocyte proliferation in liver homeostasis and regeneration. Cell Stem Cell. https://doi.org/10.1016/j.stem.2019.11.001
4. Greider CW, Blackburn EH (1985) Identification of specific telomere terminal transferase activity in Tetrahymena extracts. Cell 43(2 Pt 1):405–413
5. Bavestrello G et al (1992) Bi-directional conversión in *Turritopsis nutricula* (Hydrozoa). Aspects Hydrozoan Biol Sci Marina 56(2–3):137–140
6. Piraino S et al (1996) Reversing the life cycle: medusae transforming into polyps and cell transdifferentiation in *Turritopsis nutricula* (Cnidaris, Hydrozoa). Biol Bull 190(3):302–312

5

How Do Tissues Regenerate After an Injury?

Summary Tissues and organs naturally regenerate through the processes called tissue and organ homeostatic regeneration or physiological regeneration. We also know that different tissues undergo homeostatic regeneration at different speeds and can be classified as having faster, medium, or slower regeneration rates. Now, the question is what happens to tissues' capacity for regeneration after tissue injury (trauma). This kind of posttraumatic regeneration is called pathological regeneration, and it will depend on each animal in particular as well as each tissue type of how it regenerates or not by forming scar tissue.

As a second labor he [the king Eurystheus] ordered him [Hercules] to kill the Lernaean hydra. That creature, bred in the swamp of Lerna, used to go forth into the plain and ravage both the cattle and the country. Now the hydra had a huge body, with nine heads, eight mortal, but the middle one immortal. So, mounting a chariot driven by Iolaus, he came to Lerna, and having halted his horses, he discovered the hydra on a hill beside the springs of the Amymone, where was its den. By pelting it with fiery shafts he forced it to come out, and in the act of doing so he seized and held it fast. But the hydra wound itself about one of his feet and clung to him. Nor could he affect anything by smashing its heads with his club, for as fast as one head was smashed there grew up two. A huge crab also came to the help of the hydra by biting his foot. So he killed it, and in his turn called for help on Iolaus who, by setting fire to a piece of the neighboring wood and burning the roots of the heads with the brands, prevented them from sprouting. Having thus got the better of the sprouting heads, he chopped off the immortal head, and buried it, and put a heavy rock on it, beside the road that leads through Lerma to Elaeus.
—Apollodorus, *Story of Hercules and the Lernean Hydra (Apollodorus, The Library, with an English Translation by Sir James George Frazer, F.B.A., F.R.S. in 2 Volumes. Cambridge, MA, Harvard University Press; London, William Heinemann Ltd. 1921)*

Physiological Versus Pathological Regeneration

Although we don't realize that during everyday life our tissues and organs are constantly replenishing their old, unfunctional, or dead cells or what is called physiological regeneration or tissue homeostasis, we know this is fact because our hair and nails are constantly growing, but we don't normally think that this is going on in every single tissue of our body, faster or slower, but happening. Moreover, every time we have small injuries like paper cuts, small skin hematomas caused by moderate hits, skin blisters, or some superficial scratches, the tissue, independent of how long it takes, regenerates acquiring the same aspect, properties, and function as initially. This process is called pathological regeneration[1] but not necessarily means that the affected tissue will be able to engage in full regeneration process, as normally happens in the few cases mentioned above. Why? Well, because each tissue and organ has a limited regenerative capacity, which is normally associated with how large is the injury. In other words, when a particular tissue—which normally regenerates at certain rate—suffers damage that causes it to lose some of its cell mass, its ability to replace all those missing cells will directly depend on how fast it can naturally promote cell proliferation, differentiation, as well as recovery of tissue structure and function. If the speed needed to regenerate is faster than its capacity, then, in general terms, the tissue might end forming a scar, which is a repair process that solves partially the problem. The tissue at the scar zone does not present identical structural and functional characteristics as before the injury. In general, scar tissues are very amorphous in aspect, stiffer and less elastic than normal tissues. Of course, the pathological regenerative capacity of a particular tissue—or organ—will depend on many factors including the type of organism, age (embryonic, juvenile, or adult stage), health state, or genetic predisposition.

In principle, it is considered that the tissues that regenerate best are those that normally have faster physiological regeneration rates (homeostasis), but the efficiency in the process might differ from tissue to tissue. For instance, in humans, the skin epidermis, which naturally undergoes homeostatic regeneration every 2–4 weeks, is able to completely regenerate its tissue functions and structures after injury in a relatively short time, depending on how large is the tissue damage (Fig. 5.1, *Skin regeneration or scar formation after injury*). During the epidermis pathological regeneration process, the system may

---

[1] Tissue pathological regeneration is defined as a tissue's capacity to acquire the same structure and functionality it had before an injury. If the tissue doesn't achieve full regeneration and it only partially recovers its original structure and functionality, then the process is called repair. Normally, tissues that repair instead of regenerating display scar formation.

activate the production of more progenitor cells than normal at the expense of promoting the proliferation of basal cells (epithelial stem cells). In other words, because the injury itself may involve a loss of basal cells, the remaining basal cells at the side of the injury will engage in what is called symmetric cell division to produce more of them. Then, these new basal cells will migrate toward the basal membrane (see Fig. 5.1a). Once in place, these basal cells will engage in classical asymmetric cell division to produce a progenitor cell and another basal cell. In this way, the tissue maintains always a stem cell at the basal membrane while producing a progenitor cell. It is the progenitor cell that would start a series of cell divisions, expanding and generating the rest of the cells that form the epidermis (stratum spinosum, stratum granulosum, and stratum corneum). In other words, regeneration is driven the precise

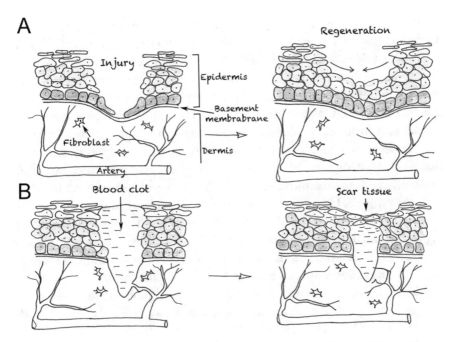

Fig. 5.1 Skin regeneration or scar formation after injury. **(a)** An epidermal injury causes tissue loss until the basal membrane has time to regenerate by activating basal epithelial cell proliferation and migration and, subsequently, progenitor cell proliferation and differentiation to generate all the stratified layers of the epidermis. **(b)** Dermal-epidermal injury does not end in normal tissue regeneration, but rather in scar formation. After the injury, blood vessels bleed to form a clot plug and cause inflammation. The clot is removed relatively rapidly; local dermal fibroblasts speed the process up by helping to rebuild the injured tissue and secreting collagen faster than normal, which generates a structure that is difficult to remodel. The tissue ends up with a repaired—not regenerated—structure called scar tissue

coordination progenitor cell proliferation and progressive differentiation until the tissue structure and functions are totally recovered.

Importantly, we know that injured areas of skin epithelium that undergo regeneration age a little more than the rest of tissue around them, which has remained at the normal, homeostatic rate of regeneration. Essentially, our original estimations about the number of cells that each tissue will produce over a lifespan (see Chap. 2) did not consider the extra number of cells that tissues may have to produce as a result of injuries. Moreover, because basal cell will eventually engage in symmetric cell division, this might promote faster aging of this particular niche. We did not account for this variable in Chap. 2 because it entirely depends on the frequency and nature of the accidents an individual suffer in the course of their life. Nevertheless, this variable can be very large and significant in certain cases, especially in burnt victims, people highly exposed to radiations (like UV light), and individual exposed to a lot of potential injuries, as is the case for construction workers, carpenters, mechanics, and others who directly expose their bodies at work.

The skin's dermis, the tissue just below the epidermis, does not regenerate as well and easily as its neighbor tissue [1]. It can often recover completely from small injuries like paper cuts (which are only 50–100 μm thick), but it usually cannot fully regenerate after larger injuries such as knife wounds or serious cuts. Instead, the dermis initiates repair mechanisms, which form a scar. Briefly, this occurs because the dermis (unlike the epidermis) is vascularized—that is, it contains blood vessels. A deep cut generates bleeding, which eventually leads to inflammation and the formation of a blood clot plug at the site of the wound. Cells from the dermis called fibroblasts start to remodel the fibrin structure around the blood clot very quickly while simultaneously trying to deposit an extracellular matrix (mainly collagen) into the tissue to reconstruct it. The problem in part is that the speed at which the fibroblasts must deposit collagen to heal the wound is faster than the normal speed they require to do a good job (i.e., a very well-organized job), so they end up producing a collapsed, disordered matrix at the site of repair. In addition, the matrix forms a keloid structure, which is hard to remodel. This visible structure is what we know as a scar and is the result of a very effective process since it is a good and fast way to solve an injured problem. Hence, it is evident that during evolution, at least in humans and larger mammals, scar formation was selected as a mechanism to heal fast, preventing infections or excessive bleedings. Some scars take years to remodel, while others never disappear. Whatever the case, scar formation is probably better in some cases than undergoing into a full regeneration process—that would take longer to accomplish—and, as a consequence, impeding or restricting mobility, which would turn the individual to be a better prey. Ultimately, it is clear that the evolutionary choice between scaring (quick plug) and full regeneration (longer attempt to rebuild the tissue structure properly) is related to the trade-off of each strategy, which varies according to the particular adaptation of each species.

One of the other tissues discussed in Chap. 4 is the hematopoietic tissue present in the bone marrow—that is, blood cells. This particular tissue can regenerate under any circumstances, like the skin's epidermis. For instance, when someone donates blood, they lose about 0.4–0.5 liters of blood (an extensive cell mass), but these cells will regenerate completely within a few weeks. In theory, a person can donate blood every 3 months, because the hematopoietic tissue will constantly produce the extra blood necessary without a problem. The same can also be extended to bleeding cases where, depending on how much blood has been lost, the patient would be able—or not—to recover after hematopoiesis. In effect, hematopoietic stem cells are capable of generating huge amounts of functional cells through homeostatic tissue regeneration (see Fig. 4.2).

In terms of pathological regeneration strategies, one of the most fascinating organs without doubt is the liver [2]. Its regenerative prowess has been known for millennia; it was even represented in Greek mythology: after Titan Prometheus stole fire from Mount Olympus to give it to humans, disobeying Zeus instructions, he was sentenced to eternal punishment for his offence. The Titan was tied to a rock, and everyday Zeus sent an eagle to eat his liver. Every night, Prometheus's liver would grow back, only to be eaten again the following day. Whether we are talking about a mythological Titan's liver or a human's, we know by now that the liver's homeostatic regenerative capacity is so potent because of the hepatocytes, which can divide and proliferate in any zone of the liver whenever necessary, especially after injury.

Of course, a human's liver takes longer to regenerate than Prometheus's (it does not quite recover overnight). But the question is: How does the liver recover after it is injured? In cases of acute mild damage (hepatocyte loss from a one-time injury), nondamaged neighbor hepatocytes undergo cell division replacing the lost ones. Interestingly, if the injury occurs repeatedly—that is, if it becomes chronic—then the regenerative process engages hepatocytes from all the zones in the liver. In severe chronic injuries, it is believed that the liver engages in the recruitment of local liver progenitor cells (LPCs) to assist regeneration. Although every major study performed with animal models (i.e., mice and rats) indicates that regeneration is mainly carried out by hepatocytes, it is also true that researchers have found that LPCs might be involved. In particular, it is not yet clearly determined whether or not LPCs are involved during human liver regeneration. More research is needed to understand the exact mechanism of liver regeneration in human pathological situations.

The liver's unique regeneration strategy may be a consequence of the fact that hepatocytes are constantly exposed to toxins (which come from the cells' fast metabolism, as well as from external sources, like ingestion). In spite of these toxins, the liver is considered one of the most regenerative organs/tissues in the body, and it may be that the organ's emphasis on hepatocyte proliferation instead of putative stem cells evolved through natural selection to promote regeneration at any cost. Throughout their evolutionary history, mammals—and especially herbivores and omnivores—have been exposed to dramatic dietary changes due to habitat transformation and large migratory events. As their diets changed, their intake began to include toxic xenobiotic compounds that the mammals' livers needed to metabolize rapidly to prevent organ injury and death. Naturally, the survivors of this natural selection process ended up with livers that had tremendous regenerative capacities, which means that today's animals (including humans) obtained this trait during evolution.

Most people think of bone as rigid and static, but it is actually a very dynamic and active tissue. Although its homeostatic replacement (physiological regeneration) process takes about 8–10 years, it regenerates well after injuries (i.e., fractures). Most clinicians who have treated patients who have suffered traumatic accidents know that if the patient has a bone fracture, the most important thing is to put the pieces back in their original place to promote normal healing. Then, as long as the distance between the fractured bone pieces is correct, the regeneration process will begin [3]. First, blood from nearby capillaries and bone marrow forms a clot mesh (called a hematoma) full of different cells, including mesenchymal stem cells and connective tissue cells. These cells start the remodeling process at the site of the fracture. At the same time, the immune system initiates inflammation, which is an essential part of the healing process. Then, within about 8 days, a soft cartilage called the callus forms between the bone ends. Over the course of 3–4 weeks, a spongy bone callus develops to replace this cartilaginous callus. The bone callus is more rigid, but it is still a transitional structure that connects the two fractured pieces of bone (Fig. 5.2, *Bone regeneration after a fracture*). Bone callus is then slowly replaced by normal bone in a process that can take several months, depending on the fracture size, until the bone totally regains its original structure and function and achieves regeneration.

Paradoxically, the tissue most closely associated with the bones, the cartilage—which is located at the joints, between the vertebras, and at the nose and ears—does not regenerate well [4]. It is actually considered probably the most non-regenerative tissue, at least, in mammals. Now, there are different types of cartilage depending basically on the properties and function. Hyaline

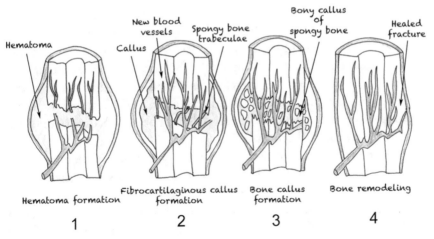

Fig. 5.2 Bone regeneration after a fracture. When bones fracture, the bleeding creates a hematoma—an early fibrin scaffold that detains the bleeding and captures local connective tissue cells and mesenchymal stem cells from the bone marrow (1). During the next week, the stem cells differentiate into chondrocytes to form a cartilaginous callus that replaces the hematoma. This callus structure also gains a new vascular network from preexisting vessels through a process called angiogenesis (2). The cartilage callus is replaced in 3–4 weeks by a sponge-like bone callus tissue that is generated after chondrocytes differentiate into osteoblasts and later osteocytes, changing the composition of the matrix to generate a stiffer biomineral formed by calcium phosphate crystals in combination with specific bone proteins including osteocalcin, bone sialoprotein, and collagen type I (3). The bone callus is slowly replaced by normal bone in a process that can take several months (4)

cartilage is the most common and is present in the joints, tip of the ribs, nose, larynx, and the rings of the trachea. This cartilage, as described below, is used during development as a precursor tissue of many bones. Probably, it is the most studied type of cartilage in terms of its biology, biophysics, and biomechanics. These special cartilage's main functions are to reduce friction at the edges of the bones and work as a shock absorber when a mechanical load is applied to the bones; hence, it is very elastic, like a rubber pad. It contains large amounts of a particular extracellular collagenous material and a special molecule called aggrecan, which gives it a unique elasticity. Aggrecan is a supramolecular structure composed of thousands of glycoproteins (polysaccharides linked to proteins), which carry a strong negative charge on the polysaccharide side that causes electrostatic repulsion between the molecules. It is this molecular repulsion that confers elasticity onto the tissue.

Fibrocartilage is found in the intervertebral discs, meniscus, and ligaments and is mainly formed of collagen fibers and proteoglycans and is also a rubber-like kind of material, but its structural features are different from hyaline cartilage. Intervertebral disc connects the vertebras like an elastic glue. In fact, it is formed by a strong ring-like structure situated between the vertebras, called "annulus," formed by very tight concentrically oriented collagen fibers that surround a central zone composed of a more viscous material, called "nucleus pulposus." The disc, therefore, is a kind of circular cushion where each vertebra sits. Finally, elastic cartilage is found in the external ear, larynx, and epiglottis and is composed of many elastic fibers that contribute to their very flexible mechanical properties.

It is not exactly known why cartilage tissue type in general does not regenerate well after injury, although there are many hypotheses. Cartilage is very peculiar tissue. Like the skin epidermis, it is an avascular tissue, but unlike the cells in the epidermis, its resident cells, the chondrocytes, have a very low proliferative capacity both during homeostatic regeneration and after an injury. One reason could be that although both tissues, epidermis and cartilage, are avascular, their particular thickness is very different. While skin epidermis thickness could be between 0.1 mm (eyelids) and 1.5 mm (in hands and feet), the average cartilage thickness could be from 1.5–2.5 mm (knee articular cartilage) to 8–10 mm (intervertebral disk). These differences in thickness definitely influence the diffusion rate of nutrients, cytokines, growth factors, as well as gas and toxin exchange that, in addition, could affect their regenerative capacity. Moreover, epidermis tissue does not suffer from low oxygen pressure levels, and therefore, their respiration gases (O_2 and CO_2) can exchange relatively fast. Instead, cartilage oxygen pressure is very low, around 0.5–1% (compared to almost 20% present in the atmosphere), and CO_2 diffusion is also slower, which for sure strongly influence in their non-regenerative phenotype. Instead, during animal development, the cartilage grow very fast, like any other tissue, but as cartilage structures become bigger, the process of osteochondral differentiation take place—in the middle part of the tissue—which is the generation of the long bone structures from previous cartilage tissue. During this process, cartilage cells, chondrocytes, differentiate first into osteoblast and then into osteocytes. Osteocytes biomineralize the extracellular matrix with calcium phosphate salts in combination with new extracellular proteins, including collagen type I, bone sialoprotein, osteocalcin, and others, forming the biomineral structure of the bone. This process makes bones stiffer and less elastic, as compared to the cartilage mechanical properties, softer and flexible. Since osteogenesis takes place in the internal piece of the tissue, cartilage remains restricted at the tip of the bones, forming the joins. Osteochondral differentiation involves also angiogenesis in the pre-bone zone, that is, the generation of blood vessels from preexistence ones, which helps to develop the biomineralized matrix of the bones. At the end of the intrauterine development and after birth (mammals), cartilage growth is progressively reduced, and joints will grow at slow rate during childhood and then slower through puberty until early adulthood. During the rest of the life, cartilage growth is almost nonexistent. Thus, adult chondrocytes are terminally differentiated cells like hepatocytes, but they cannot engage in proliferation themselves or activate resident stem cells, so injured cartilage tissue degenerates, progressively losing its mechanical properties until the joint or the intervertebral function is significantly reduced. Moreover, after injury, the tissue initially seems to engage in a regenerative response that rapidly ended and is overtaken by an inflammation reaction and scar formation. In most cases, the tissue after injury undergoes a degenerative path, which includes a sequence of events that end in the destruction of the tissue structure (degeneration), losing all the particular biological and biomechanical functions.

Therefore, most athletes who have the bad fortune to suffer a joint cartilage injury are unlikely to recover, and these injuries can permanently weaken their performance. Believe it or not, even today, there is no good therapeutic solution for cartilage trauma.

One other organ that presents almost no regenerative capacity after injury is the heart. It is well known that after the blockage of the coronary arteries, it causes a cardiac infarction—a heart attack—the heart's myocardium cells (cardiomyocytes) die of necrosis[2] because they have a low or nonexistent supply of blood. These dead cells are removed in part by immune system cells (mainly macrophages) that partially clean up the site of the infarction and start a remodeling process without replacing the lost cardiomyocytes, generating a fibrotic tissue that is stiffer than the original tissue. Thus, the local scar that forms is never removed, and it has a global effect on the rest of the heart, inducing what is called cardiac failure, in which the walls of the affected ventricle start to dilate (due mainly by the presence of nonfunctional cardiac muscle). Cardiac failure results in a progressive loss of functioning, which may even cause the patient's death. Recent studies have suggested that the human cardiac muscle does not have much intrinsic (i.e., homeostatic) regenerative capacity because there are either no or few resident stem cells present to assist in the healing process. This is one of the reasons why cardiovascular diseases are the leading cause of death worldwide. Similarly, spinal cord injury, in which the spinal nerves are disconnected from one another, is currently considered non-regenerative in mammals.

---

[2] Necrosis is defined as cell death caused by a reduction in physiological nutrient levels, gas (O_2 and CO_2) exchange, or toxin removal. Necrosis can also be caused by trauma-induced tissue injury, extreme heat, or exposure to caustic and corrosive substances.

Regeneration in Animals Different from Mammals

Regeneration, physiological or pathological, operates in different ways according to the specific animal type, but in general, invertebrates—such as worms and sea stars—present more regenerative capacity than vertebrates. Moreover, among the vertebrate group, amphibians and reptiles have better regenerative capacity than mammals. Thus, it seems that the simpler the animal is, the better it regenerates after injury. This capacity is particularly impressive when body part or limb amputation takes place. For instance, flatworms and hydras can regenerate completely after their bodies are sectioned in tiny pieces, phenomena that were probably known in the ancient Greece and, therefore, were the inspiration of the mythological story about Hercules and the immortal Hydra.[3] Sea starts can also regenerate their limbs and entire body if the central nerve ring is not harmed after amputation. Hydra tissues contain three types of stem cells: ectodermal and endodermal stem cells, which are unipotent (meaning that it can produce only one type of differentiated tissue), and a multipotent stem cell (MPSC), also called interstitial stem cell (ISC) [5]. This ISC generates the nerve cells, gland cells, nematocytes, as well as germline cells. After bisecting the hydra into two, separating the head with tentacles from the foot, a regeneration process rapidly takes place where each of the stem cells participates in an organize way to generate a new head in the foot and a new foot from the dissected head. This process can go "over and over," indicating that this incredible organism can undergo regeneration after regeneration, if necessary.

Curiously, small vertebrates like salamanders (newts and axolotls) as well as tadpoles (but not froglets or adult frogs), are capable of regenerating multiple body parts after amputation including the limbs, tail, and jaw, which is an incredible process considering the complexity of the tissues present in such structures [6]. In mammals, including mice and humans, limb regeneration after amputation is only possible at the tip of their toes and fingers, respectively, in a process that resembles the limb regeneration in salamanders [7]. Curiously, it has been also reported clinical cases of fingertip regeneration after traumatic injury in humans, which is quite surprising but nevertheless real. This capacity is progressively lost after puberty, decreasing in adults, and almost nonexistent in elderly people. Now, regardless of the animal type, salamanders, mice, or humans, the question is: How does limb regeneration proceeds? Years ago, when scientist started investigating the regenerative mechanism, it was obvious to postulate that some type of multipotent stem cell (MPSC) was present in the limb's tissues that directed the regeneration

---

[3] Apollodorus, The Library, with an English Translation by Sir James George Frazer, F.B.A., F.R.S. in 2 Volumes. Cambridge, MA, Harvard University Press; London, William Heinemann Ltd. 1921

process, as previously observed in invertebrates, like hydras. The other reason in suggesting the presence of a MPSC was because limbs, as we all know, are formed by many tissues including the bone, cartilage, muscle, fat, skin dermis, and epidermis as well as blood vessels and nerves. Therefore, each tissue uses their own stem cell niche to regenerate its resident tissue (in an orchestrated way), or a single MPSC proliferates and differentiates into every single tissue type, recapturing full shape and function. To their surprise, no specific resident MPSC nor activation of specific tissue-type stem cells was found to be responsible for such full-limb regeneration process. Instead, the limb at the injury site undergoes a process that is particularly extraordinary.

Salamander limb regeneration is probably the most studied and characterized by researchers and, therefore, is a good example to describe next (Fig. 5.3, *Salamander limb regeneration*) [6]. Thus, after limb amputation in salamander, the remaining tissue first undergoes a process of blood clot to stop bleeding—as in any regular bleeding injury. Then, the epithelial cells at the surface of the skin rapidly proliferate and migrate to cover the wound surface forming a structure called wound epidermis. This cellular layer ensures the insulation of the internal compartment, which is the tissue underneath, with the exterior environment. This step is critical since the formation of the protected epidermal layer helps to reduce the oxygen levels in the internal tissues, which promotes the next-regeneration steps to proceed. At this time, the tissue underneath the epidermal coat undergoes cell proliferation developing a ball-like structure called blastema. Blastema cells are a very peculiar type of cells, since there are not derivate from stem cells (as was mentioned above), but instead, they are the bone, cartilage, muscle, fat, and dermal cells that have dedifferentiated—losing their phenotype—and acquiring characteristics more comparable to stem cells. Then, these blastema cells proliferate to form the particular regenerative structure. This process is really amazing since we are saying that finally differentiated adult cells would undergo a process that consists in turning back into more primitive type of cells, to an early stage of development, when limb primordium appears.

Now, if we relate this event with the liver regeneration mechanism (described above), it seems that it is not very different, in principle, since hepatocytes undergo proliferation and dedifferentiation to produce two new daughter cells that after redifferentiation will turn back into finally differentiated and functional cells. Well, in the blastema, something very similar to the liver happens, though liver regeneration does not proceed by blastema formation. It is believed that each of the affected tissues contributes to generate a dedifferentiated cell that will form a blastema, but, interestingly, it seems that each blastema cell will be able to produce, after proliferation, the tissue cell type from its original lineage. In other words, blastema cells precedent from the bone will form only bone cells, and those recruited from muscle will

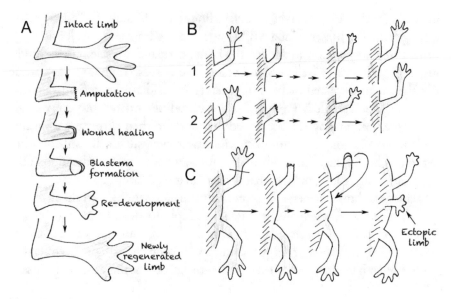

Fig. 5.3 Salamander limb regeneration. (a) After limb amputation, blood clot stops bleeding, and wound healing starts followed by epidermal tissue growth on top to cover the insured site forming a structure called wound epidermis (step 1). Tissues underneath the wound epidermis proliferate forming a blastema (step 2). Blastema cells are formed by the bone, cartilage, muscle, fat, and dermal cells that have dedifferentiated—losing their adult phenotype—and acquiring characteristics more comparable to stem cells. Next, the blastema undergoes a process called redevelopment where the regenerative structure undergoes a series of processes including proliferation, cell differentiation and morphogenesis needed to generate a new embryonic-like limb (step 3). Finally, the new limb will grow until reach the size, shape and identical function as the lost one (step 4). (b) Limb amputation in salamander will be able to form a blastema structure and recapitulate redevelopment to regrow the remaining missing part independent of the amputation site, such as (1) wrist-level amputation (that will generate a hand) or (2) shoulder-level amputation (that will generate an arm), which indicates that the blastema itself has positional information also called positional memory. (c) Blastema structure is an autonomous self-organized system that carries all the information to develop the remaining part of the limb. After taking salamander's limb blastema and transplanted ectopically—at the abdomen of the animal body—it is able to form a limb at the transplanted site

generate only different type of muscle cells (striated o smooth), but not cartilage of skin dermal or epidermal cells, and so on.

In resume, the blastema structure is like a "recruited and rejuvenator" of adult cells that "convert" them into a more primitive cell structure that undergoes next to a process called redevelopment. This incredible process basically recapitulates development. In other words, the blastema recreates all the developmental processes including proliferation, cell differentiation, and morphogenesis needed to generate a new limb embryonic-like limb. Thus, the blastema will turn into a small limb (like a baby limb) which will grow and develop into a large and adult limb (Fig. 5.3a). Even more, depending on where in the limb the amputation occurs, the blastema will be able to recapitulate redevelopment to regrow the remaining missing part, which indicates that the blastema itself has positional information also called positional memory (Fig. 5.3b). In other words, wrist-level blastema cells will regenerate a hand, and shoulder-level ones will regenerate an arm [5]. In addition, it was shown that the blastema structure is an autonomous self-organize system that carries all the information to develop the remaining part of the limb. It was clearly demonstrated after taking salamander's limb blastema and transplanted ectopically (at different position of the animal body) and showed that it was able to form a limb at the transplanted site (Fig. 5.3c).

Salamander Limb Regeneration: Step by Step

After limb amputation, a complex regeneration process takes place which is normally divided into many sequential steps (Fig. 5.4, *Salamander limb regeneration in detail*): *healing*, a process consisting in the formation of wound epidermis where the amputation site is rapidly covered by an epithelial cell layer; *dedifferentiation*, an active disorganization and dedifferentiation of mesenchymal tissues near the wound, including dermal, muscular, and bone cells that will form an undifferentiated mass of cells called blastema (The blastema is covered by an apical epidermal cap (layer).); *cone stage*, formed by the proliferation process of the blastema cells; *palette stage*, early stage in the redevelopment stage where blastema cells start to acquire morphological and early pattern formation as a primordial limb during development; *notch stage*, redevelopmental program continues and evidences of embryonic structures are visible where the main bone structures start to appear; and *digit stage*, the entire limb organization is defined where all the structures are obtained including bones, joins, muscle, dermis, and skin as well as blood vessels and nervous system. The process recapitulates development, and the regenerated limb is structurally and functionally identical to the amputated one. Moreover, the same signaling process and gene regulation mechanisms are identical as early development.

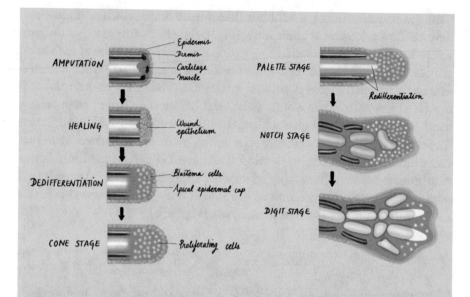

Fig. 5.4 Salamander limb regeneration in detail: after limb amputation, regeneration takes place which is a sequential process that involves the formation of a blastema structure, formed by a dedifferentiated cell mass from neighbor mesenchymal tissues, which undergo redevelopment, where early embryogenesis is recreated in an identical way as limb development. Artwork made by Claire Jarrosson Moral

Of course, for salamanders or tadpoles, tinny animals, this process could take few weeks, which is okay because salamanders naturally reach adulthood in several weeks. The animal can accomplish regeneration of a limb in a relatively short time. The final regenerated limb is exactly identical as the lost one, structurally and functionally. Moreover, if another amputation occurs, the process will repeat again as many times possible along the life of the salamander.

This indicates that, in principle, the formation of a regenerative blastema would follow the repetitive sequence described above for the liver (and also in Prometheus's tragedy, which tied to a rock, an eagle used to eat his liver every day, and every night, his liver would grow back, only to be eaten again the following day). Moreover, we can also relate this event to the process that our immortal jellyfish, *Turritopsis* sp., naturally undertakes. As previously described in Chap. 4, adult jellyfish individuals—named medusas—can opt to go through a reverse cycle (Fig. 4.6), which consists in a drastic bodily transformation to prevent aging and death. By the end of this reverse cycle, the organism has once again become an undifferentiated cyst that sticks to the seafloor. From that simple structure, a new polyp emerges to reinitiate another regular life cycle. In essence, what *Turritopsis* sp. does with its entire body salamanders do at the injury site by forming the blastema. This last concept is very important because this could mean that, by extension, even mice and humans have a regenerative capacity at the fingertips, in a very similar way to the immortal jellyfish. Even though, the question is: Why in mammals this capacity is restricted only to the digit tips? Clearly, it is not a tissue size issue since salamander's limbs and mice toes are very similar in that aspect. Thus, it is believed that the reason could be the fact that immune systems among amphibians and mammals present some important differences. Briefly, in mammals, wound healing proceeds after the formation of a blood clot followed by a wound healing process where inflammation and innate immune response take control (see Fig. 5.1b). These steps, orchestrated by the immune system, prevent potential microbial pathogens to colonize the wound (generating and infection) as well as to accelerate the closing of the tissue defect produced by the injury. In this way, this mechanism is very effective in recovering (partially) the tissue structure and function, but as a consequence, a scar tissue is developed. In other words, mammals during evolution "have selected" this wound healing process—to the detriment of producing scar tissue—in favor of being faster and efficient. Thus, scar tissue development was naturally selected than regeneration, which would take longer. Amphibians, on the other side, maintained full-regeneration capacity because in their environment a missing limb probably did not affect much of their survival capacity nor their reproductive fitness. They are also very quiet animals and are cold blooded, and as a consequence, their metabolic activity is significantly lower than mammals; in addition, they can reduce to the minimum their movement activity during the process of regeneration. Mammals instead are very active animals which have much higher metabolic activity, and their behavior is more dynamic and complex than salamanders. Therefore, to maintain a regenerative process that strictly involves a redevelopment (very fragile step) is probably out of the question. During this process for sure, the regrowth of an embryonic-like limb is probably incompatible with their multiple survival necessities (search for food, escape from predators, social activities, etc.). Then, with scar formation after limb amputation generation, a stump was probably the safer and most adaptive way to go for mammals. However, it is clear that mammals, like humans, keep a reminiscence of this astonishing way of regenerating in our fingertips, right below of our nails, and a group of cells—the same cells responsible for producing the constantly growing snail tissue—are capable or engage in a regeneration process via blastema formation and redevelopment of our fingertips. This embryonic-like fingertip regrowth seems to have no effect on our survival capacity nor our reproductive one. The question now is: Could we learn from our fingertip regenerative capacity to be able one day to extend other body parts, including entire limb regrowth?

In the last decades, a huge amount of research efforts in biomedicine have been dedicated to improving our understanding of the different regeneration mechanisms. Specially, the main effort is focused on learning from those organisms that do regenerate well some tissues and organs in contrast to humans. While the research community has advanced tremendously in this area, there is still a long way to go. This short book is designed to aware the reader about the main concepts in cell biology, developmental and evolutionary biology, that eventually could help to inspire others to look for solutions or new perspectives on therapeutic strategies to assist future therapies related to ameliorate or cure some diseases or injuries produced by trauma.

In conclusion, in this chapter, it can be stated that:

- All tissues and organs are constantly under physiological regeneration or what is called tissue homeostasis.
- When tissues and organs, independent of how long it takes, regenerate acquiring the same aspect, property, and function of normal tissues, this process is called pathological regeneration.
- If after injury tissues or organs do not acquire the same aspect, property, and function as before the injury, then this process is called repair and in most cases implies scar formation.
- In humans, the skin epidermis, liver, and bone regenerate well after injury, which indicates that it undergoes pathological regeneration.
- Other tissues and organs in humans like the skin dermis, cartilage, heart, and spinal cord do not regenerate after injury, and in most cases, it undergoes repair that ended in scar formation.
- Other animals rather than mammals in general present more regenerative capacity after injury, including invertebrates and some vertebrates.
- Invertebrates in general regenerate very well, like sea stars, flatworms, or hydras, where full recovery and function after body sectioning is accomplished.
- Small vertebrates, like amphibians (salamanders and tadpoles), regenerate well after limb amputation through blastema formation and redevelopment process. Froglets and adult frogs do not regenerate well after injury.
- Mice and humans can regenerate toe tips and fingertips, respectively, through blastema formation and redevelopment, indicating that also mammals present this amazing unique regenerative capacity.

References

1. Reinke JM, Sorg H (2012) Wound repair and regeneration. Eur Surg Res 49:35–43
2. Chen F et al (2019) Broad distribution of hepatocyte proliferation in liver homeostasis and regeneration. Cell Stem Cell. https://doi.org/10.1016/j.stem.2019.11.001
3. Ferguson C et al (1999) Does adult fracture repair recapitulate embryonic skeletal formation? Mech Dev 87:57–66
4. Tuan RS et al (2013) Cartilage regeneration. J Am Acad Orthop Surg 21(5):303–311
5. Galliot B (2005) Regeneration in hydra. Encyclopedia of life sciences. Wiley. www.els.net
6. Joven A, Elewa A, Simon A (2019) Model system for regeneration: salamanders. Development 146(14):dev167700
7. Seifert AW, Muneoka K (2018) The blastema and epimorphic regeneration in mammals. Dev Biol 433(2):190–199

6

Epilogue: The Future of Humankind Could Depend on Unicellular Life

Summary It is very clear for most of us that life in our planet is becoming more and more restricted in terms of having enough resources to support everyone's subsistence. This without question could cause in the near future a massive crisis for survival. But before this would happen, we could have one or two more chances to survive. Our first option is to solve our population growth dynamics, and the second is to prepare ourselves for a long trip to other worlds, outside our solar system. Both scenarios are considered to evaluate how we can contribute to the never-ending story of life.

> *Is it absurd to imagine that our social behavior, from amoeba to man, is also planned and dictated, from stored Information, by the cells? And that the time has come for men to be entrusted with the task, through heroic efforts, of bringing life to other worlds?*
> —Albert Claude, *From Nobel Prize Lecture (Dec. 1974), "The Coming Age of the Cell." Collected in Jan Lindsen (ed.) 1992*

We humans have undoubtedly already reached the maximum population that our planet can support. We do not have enough resources to maintain our own lives nor other species' lives. In population biology, this is called the point of inflexion: the moment at which a population becomes unsustainable due to a shortage of resources. When a species reaches an inflection point, a significant, drastic change happens—usually, a severe reduction in the population or, in the worst-case scenario, extinction. In our case, both outcomes are possible, but the first is likely the best solution we have: by gradually having fewer children and reducing our natality rate, we might after several decades be able to maintain or reduce our population to reach an equilibrium with nature. This is the most desirable near future scenario for human life on Earth, but many people are thinking about the possibility of colonizing new worlds outside our own. This is the driving force for current explorations of planets where life may be able to continue (human life and probably many animals, plants, fungi, bacteria, and—why not—some archaea). We certainly hope that the second possibility (extinction) will not occur, but in the event that it seems imminent, there is some thought that humankind's primary rescue strategy—the modern Noah's Ark—may have the planet Mars as its final destination: our new world.

Mars is relatively close compared to other potential planets outside our solar system (exoplanets), but it is clearly not an ideal destination. It is very cold and does not have liquid water, which is a major problem, because liquid water is an essential factor for sustaining life. In addition, its atmosphere has only small traces of oxygen compared to the Earth's (which is 21% oxygen); instead, Mars's atmosphere is composed of 95% carbon dioxide (CO_2), 2.6% molecular nitrogen (N_2), and about 2% argon (Ar). In addition, Mars is buffeted by frequent, strong windstorms and is almost entirely covered in red dust (which is mainly clay). Lastly, the red planet is much smaller than the Earth, with a diameter that is only about half as long, which means that the gravitational force operating on the planet's surface is also weaker—(3.7 m/s^2)—than on Earth (9.8 m/s^2). Under these altered gravitational circumstances, humans (and most vertebrates) would undoubtedly suffer health problems, including the loss of bone structure and muscle atrophy. Nevertheless, three independent space programs (NASA, SpaceX, and Boeing) are actively competing to be the first to establish a human colony outside Earth but still into our solar system (Mars).

An alternative to Mars is presented by the growing exoplanetary discovery effort (which is led mainly by the NASA Exoplanet Project). Researchers have detected more than 400 potential exoplanets that contain liquid water, but many of them are much bigger or much smaller than Earth. Moreover, even with this large catalogue of possible habitable exoplanets that humans might one day colonize, those that offer similar conditions to Earth are extremely far away—several light-years (LY) away, in fact.[1] The closest known exoplanet is *Proxima Centauri b*, which is about 4.24 LY away, orbiting the habitable zone of the red dwarf star *Proxima Centauri*. The second closest is the planet *Gliese*

---

[1] A light-year distance is defined as the number of kilometers (or miles) that light can travel in 1 year—a huge distance. Since the speed of light in space is about 300,000 km/s, light travels about 9500,000,000,000 km (9500 billon km) in a year.

667 Cc, which orbits the habitable zone of the red dwarf star *Gliese 667 C*, about 23.6 LY away. The third, *KOI-4878.01*, is *very* far away (1135 LY), within the habitable zone of the yellow star *KOI-4878*. All these exoplanets are very similar to Earth, but *KOI-4878.01* is considered Earth's "twin planet" and appears to be almost identical. Its Earth Similarity Index (ESI) is around 98%, because its mass is almost exactly the same as the Earth's (0.99); its median temperature is 16 °C (compared to 15 °C), which suggests that the planet's water will be liquid; and it takes about 449 days to rotate around its star, so its calendar year is only a little longer than ours. The only problem with the planet is how far it is from Earth. On the other hand, *Proxima Centauri b*'s median temperature is much lower (-39 °C), and because of its proximity to its star, it is "tidally locked" like Earth's moon, by the strong gravitation force generated by it star, which means that the planet does not rotate, and the same side always face the star. One face, then, is very hot, and the other one is very cold, so the habitable zone lies at the border between perpetual light and perpetual dark. It is estimated that in this intermediate zone—the penumbra— the temperature would be around 0–1 °C, which is fairly cold anyway. Besides, the absence of day and night could be kind of disturbing, at least for humans and most of the animals and plants, who are adapted to circadian rhythms[2] on Earth. Our third potential new home, *Giese 667 Cc*, has an ESI of 84%, with a median temperature of 13 °C and a gravitational force 1.3 times stronger than the Earth's, but it may be also tidally locked to its star. Worse, the side of the planet that faces the star is constantly bombarded by X-rays and heavy ultraviolet (UV) radiation, which have created an environment that is not appropriate for life—at least not the kind of life we know.

We could continue to consider more and more exoplanets, detailing their pros and cons, but ultimately, we would be forced to travel to the closest one, regardless of whether it is the most ideal for sustaining human life or not. The reason we would have to make this decision is simple: 1 LY is a tremendous distance that would take several thousand years for a spacecraft to cross, even if it could travel at a very high speed. For instance, if we decided to go to *Proxima Centauri b* (4.24 LY away), a spacecraft equipped to travel at about 300,000 km/h would take around 15,000 years to reach the planet. Alternatively, if we endeavored to go to the most distant and most hospitable of the planets we've discussed, *KOI-4878.01*, a spaceship traveling at the same speed would take 4 million years to arrive. Our present technology could not

---

[2] Circadian rhythms are physiological, mental, and behavioral patterns generated as a consequence of the transition from day to night, which are mainly regulated by the presence of light or dark. The fact that most animals are awake during the day and asleep at night is a clear consequence of the presence and absence of light, respectively, at those times.

possibly support such a long voyage. The 15 closest known exoplanets are all fewer than 30 LY away (including *Gliese 667 Cc*, which is 23.6 LY from Earth), and this is almost certainly the absolute farthest we could travel, unless there is a dramatic new breakthrough, like faster spacecraft or the ability to travel using spatiotemporal tunnels or Einstein-Rosen bridges (also called wormholes[3]). For now, these possibilities are still out of question.

Regardless of whether it is possible to create a wormhole and whether our destination is Mars or *Proxima Centauri b*, it is clear that any spacecraft that could carry humans across these immense distances would have to be controlled by robots while the human inhabitants were maintained by some kind of low-temperature hibernation system (which has not yet been developed). For instance, a trip to Mars is expected to take around 1.5–2 years, which is a long time even for expert astronauts. Moreover, once they have landed, these astronauts would have to live inside the pressurized ship module for a while until they were conditioned to build a colony that would be comfortable enough to offer a decent quality of life. It would take some time—maybe years—to send them all the infrastructure they would need to establish a functional and healthy colony. Given that this is all very challenging, an alternative strategy could be to send a few humans, to establish a sound and stable colony, and to send the rest of the planet's future inhabitants as unicellular human equivalents: that is, as gametes, zygotes, or early embryos. These could be easily frozen in liquid nitrogen (N_2) at -180 °C and maintained for a long time—hopefully for 15,000 years, in case we change our plan and decide to colonize (for instance, *Proxima Centauri b*). In this way, humans (as well as other organisms we would need to populate our new world, including animals, plants, fungi, and bacteria) could be stored (along with all their genetic variability) in relatively small compartments aboard a spacecraft. This is relatively straightforward and looks very simple to do. With the technology we have today, this project is feasible, though it would of course be complex. To maximize the chances of success, multiple starships could be sent simultaneously to Mars and to different exoplanets.

The question is, once we have landed on our new planet(s), how would we take these cells and early embryos out of their frozen state and usher them through the next stages of their development? If the planet in question is Mars, the chosen group of astronauts would need to be very careful at this

---

[3] A wormhole is essentially a speculative structure that would hypothetically connect two distant points in space. It is based on a specific solution of Einstein's field equations. If a future civilization is able to create a wormhole, the wormhole could be used for interstellar travels. See Morris, Michael S. and Thorne, Kip S. and Yurtsever, and Ulvi (1988). *Wormholes, time machines, and the weak energy condition.* Physical Review Letters, 61 (13). pp. 1446–1449. ISSN 0031-9007.

juncture. Now, for most of the inhabitants of these hypothetical Noah's Ark starships, development should not pose a problem, since most free-life development animals (fishes, amphibians, birds, reptiles) as well as insects, plants, fungi, and bacteria will undergo development, reach adulthood, and pass life on to the next generation without much assistance as long as they are in an appropriate, controlled environment. As long as we want to populate our new world *without* mammals, the project seems feasible with our present technology. Then, once the first astronauts have replicated an environment that is reasonably similar to Earth's on the new planet, massive immigration could happen easily. More ships would transport more people (colonists) to the new world, just as they did throughout the long history of colonization on Earth.

The scenario looks very different, however, if our target is *Proxima Centauri b* instead of Mars. It is unlikely that any of us will witness such a migration, mainly because we don't know how to keep humans under hibernation for 15,000 years. Even if we develop a spacecraft that would travel much faster, let's see, 100 times faster than our previous one (300,000 km/h), it would take around 150 years, which is still a long time anyway.

If the mission's goal were to send mammals, therefore, the challenge would be to build and send a machine-based system that could first replicate maternal intrauterine life and then take care of mammals (including humans) after they were born. The system would need to be capable of supporting our growth, maturation, and education until we were old enough to be independent of this "artificial mother" and ensure the beginning of a new exoplanetary generation. This is science fiction, of course, but as we know by now, many ideas that begin as science fiction turn into real human achievements. The question is how we will envision these difficult tasks from a scientific point of view, and the answer is complex, because it would be a tremendous challenge to create the equivalent of a maternal intrauterine environment that would ensure that mammals could undergo a physiological development *in-machine*. In order to build such a "mother equivalent," we would need to be able to build a uterus-like containing machine from scratch. Thus, all the tissue cells or stem cells (i.e., uterus epithelial cells, connective tissue cells, vascular endothelial cells, etc.) necessary to build the *in-machine mother* would need to be stored in the same way as the gametes, zygotes, and early embryos, and the in-machine mother would later foster. This device—a complex bioreactor that would be capable of supporting mammal development—would need to be built with all the necessary requirements to ensure blastocyst attachment to the *in-machine* uterus epithelium and placenta formation (from a hybrid tissue composed of extraembryonic tissue and *in-machine* tissues). Moreover, the artificial placenta and umbilical cord would need to grow in tandem with the embryo-fetus to meet its demand for nutrients, toxin excretion, and gas exchange (O_2 and CO_2), just as real mothers do. In other words, the machine would need to meet all the necessary requirements to ensure that development happens in the most natural way possible during the 9-month period of intrauterine development (for humans). This device, which would need to be developed and tested on Earth, could be a remarkable interdisciplinary scientific project; cell biologists, developmental biologists, tissue engineers, bioengineers, and physicians could work together to ensure that it achieves its assigned task. The device would need to be ready to assemble itself on the starship, just after the ship arrives on the new planet and confirms easily by sending a probe that it has all the necessary elements to sustain life as we know it on Earth. In this way, the first step could be accomplished. The first question at this time is how ethic it is to "test" these devices with human embryos before we decide to send it to a long trip. It seems very clear that it would not be generally accepted by an ethical committee. But in the event that we could do testing with animal models that ensure that humans would grow and develop well, we are postponing the ethical problem into the future. This is a really weak part of the mother equivalent project, as expected.

Next, we need to consider how the colonizing platform's program would take care of its newborns after intrauterine development is complete. Once again, this task would have to be performed by artificial intelligence units (AIUs), which would take care of the initial generation of humans. Then, once this first generation reaches maturity, the new humans could gradually start helping the AIUs develop a second group of frozen embryos. These first humans would start performing tasks that the machines are incapable of, establishing basic human-to-human contact, affection, and a sense of family and group work, solidarity. As time goes on, the colony of humans would grow and become more and more independent from the AIUs. During this stage, the machines would still perform essential tasks, like preserving additional stored gametes, cells, and embryos from every species to ensure that if the colony needs to be restarted or restored after a setback for any reason, there are enough reserves for that.

At this point, an interesting paradox is generated which is the fact that *we*, humans—as well as many other species—will become the invaders of a new world. If we remember what we have learned from the book by H. G. Wells (*The War of the Worlds*) about the invaders, who arrived to our Earth with the only purpose of destroying and dominating every single living thing, using the most devastating and effective weapons, leaving humans almost at the edge of their extinction, then, the horrible and terrible invaders of another world could become ourselves. Why? Because most probably, since we are selecting "ideal" planets to sustain our life, it is highly probably that some kinds of life already exist at the time we arrive to our new "home." If this is the case, most probably the life that has originated at this planet is totally or partially noncompatible with our own. This is because independent of the kind of bacteria, archaea, or eukaryotic cell, multicellular organisms living there most probably are highly adapted to their environment and we, as intruders, devoid of any possible adaptation strategy, would become their target, to parasite, to be infected, killed, and eaten. This means that we would be arriving to a world that would be populated of living forms that would become a potential threat to our mission. The next decision will be straightforward: to annihilate any form of life of this planet to ensure the success of our landing life-forms. The question is: Do we have the right of doing such thing? Moreover, is it ethically accepted? Can we go anywhere in the universe colonizing, submitting, or destroying other forms of life just because we decide just to do it, without thinking its consequences? Could we cease and entire process of evolution of new life-forms just because we don't know how to manage our own life in our planet? Some kind of interplanetary law might logically impede a life-form to just extinct another, right? After all, this was the perfect argument used in Wells' book to fight against the invaders from another world. If works in his book it must work in a potential future situation. This, from my point of view, regardless if we are able to accomplish such exoplanet colonization, is an important point to take into consideration at the time we decide to engage in such exoplanetary colonization program. Moreover, another problem to be solved, at least in the case of Mar's colonization, is the fact that cargo spacecraft carrying merchandise (such as valuable minerals) will have to travel back to Earth. This simply implies that if these spacecraft are not perfectly decontaminated, the danger of transporting new potential pathogens from the red planet is high. Since, most probably these new forms of life cannot be combated with our own medicines, the risk of trigger an unstopped pandemic is eminent. This indeed would turn into a real *war of the worlds* on Earth.

We don't have any rights to perturb or annihilate any form of life in our planet or in any other world in the universe. We have to be very conscious about our next steps because we don't have many options to ensure the continuation of, at least, many species' life on Earth, including humans. I am sure that life will find their way, in different forms, like what happened before during Earth's history after many species' extinctions during evolution. Without doubt, many unicellular as well as multicellular organisms will survive our civilization, even if we are destroyed by our own weapons. New species will appear and evolve into complex systems and societies. The question is if we are ready to do it for the present species and ourselves.

If we learn how to take care of our Earth and the millions of other species that share the planet with us, the exoplanetary adventure might remain a tale, a science fiction adventure. On the other hand, if we persist in our destructive and abusive behavior, this fiction will become more and more real, and we would have to deal with the paradox of being the new world invaders. If we do not change the way we treat our planet, the never-ending story of life—a story that might be very common in our universe—could produce our planet life extinction.

Glossary

Adenosine-5-triphosphate (ATP) The most common nucleotide used to store or release free energy in cells.

Adaptive evolution Refers to the increase in frequency of beneficial alleles and decrease in non-beneficial ones due to natural selection.

Aerobic A cell or organism that utilizes O_2 or that can grow in the presence of O_2.

Anaerobic A cell or organism that functions in the absence of O_2.

Apoptosis Also called programmed cell death, is a well-defined process leading to cell death regulated by the own cell.

Archaea Class of microorganisms without nucleus and organelles that constitutes one of the three distinct evolutionary lineages of modern-day organisms. Also called *archaebacteria* and *archaea*

Bacteria Class of microorganisms without nucleus and organelles that constitutes one of the three distinct evolutionary lineages of modern-day organisms.

Base pair Is the association in ADN or RNA of two complementary nucleotides linked by hydrogen bonds at the base component of each of the molecules involved. In ADN, adenine pairs with thymine (A-T) and guanine with cytosine (G-C). In RNA, adenine pairs with uracil (A-U) and guanine with cytosine (G-C).

Blastula An early embryonic form produced by cleavage of a fertilized ovum and usually consisting of a single layer of cells surrounding a fluid-filled spherical cavity.

Cell cycle Sequence of processes in which a cell first duplicates its genetic material (chromosome/chromosomes) and then divides into two new cells or daughter cells (cell division) carrying the same amount of genetic material as the original cell.

Cell division Process during the cell cycles after the cell duplicated its genetic material (chromosome/chromosomes) consisting in forming two new cells or daughter cells carrying the same amount of genetic material as the original cell.

Cellular evolution Refers to the cellular origin and consequently evolutionary process of cells (adaptation, natural selection, diversification).

C. E. Semino, *The Never-Ending Story of Life*, https://doi.org/10.1007/978-3-030-75969-8

Cell wall A rigid extracellular structure deposited outside of the plasma membrane (cytoplasmic membrane) mechanically protecting the cell. It is present in plant, fungi, bacteria, and archaea but not in most animal cells (mainly multicellular organism).

Chloroplast A specialized membrane-limited organelle presented in vegetal cells containing chlorophyll-associated structures used to capture light for photosynthesis.

Chromatid One copy of a duplicated chromosome, formed after DNA replication, that is still attached by the centromere to the other chromatid.

Chromatin Complex of DNA and histones (and other nonhistone proteins) forming the eukaryotic chromosomes.

Chromosome In eukaryotes, is the unit of a genetic material consisting in a linear double-stranded DNA molecule and their associated proteins (histones and non-histone proteins). In bacteria and archaea, a single, circular double-stranded DNA molecule.

Chronocyte A hypothetical predecessor of eukaryotic cells during evolution that engulfed heterotrophic bacteria and archaea to generate the first animal and fungi cells and later engulfed cyanobacteria to generate the first vegetable cells.

Collagen A family of long linear triple-helical type of protein which is present in most extracellular matrix (ECM) of animal cells.

Crossing over Exchange of genetic material between maternal and paternal chromatids during meiosis producing recombined chromosomes.

Cyanobacteria A group of photosynthetic bacteria.

Cytoplasm Viscous contents of a cell that are contained within the plasma membrane in bacteria and archaea. In eukaryotes, between the plasma membrane and the nucleus.

Cytoskeleton Network of fibrillar structures (microtubules and microfilaments) present in the cytoplasm of eukaryotic cells that provides mechanical structure as well as cell movements.

Deoxyribonucleic acid (DNA) Long linear double-helix polymer composed of four types of deoxyribose nucleotides which is the carrier of genetic information.

Development General process involving cell proliferation, migration, and differentiation by which a zygote gives rise to an adult individual.

Differentiation Cellular process by which a progenitor precursor cell becomes a mature specialized cell type.

Diploid An organism or cell having two homologous chromosomes.

Ectoderm Together with the endoderm and mesoderm is one of the three primary cell layers of the animal embryo and contributes to the formation of epidermal tissues, the nervous system, and external sense organs.

Embryogenesis Early developmental process of an organism from zygote to embryo states.

Embryonic stem cells Cells isolated from the inner cell mass of a blastocyst that can be expanded by culturing in vitro maintaining its pluripotent capacity (able to differentiate in many cell types).

Endocytosis A process of invagination of the plasma membrane in order to internalize extracellular materials.

Endoderm Together with the ectoderm and mesoderm is one of the three primary cell layers of the animal embryo and contributes to the formation of all gastrointestinal organs, the liver and pancreas.

Endoplasmic reticulum (ER) Two types of membranous structures present in the eukaryotic cytoplasm with different associated functions: (1) rough ER which contains ribosomes and is dedicated to synthesize and secrete proteins and (2) smooth ER which lacks ribosomes and is in charge of synthesize lipids.

Enzyme A biological molecule with catalytic activity.

Epithelium Type of tissue present at the outer surface of some organs (i.e., gastrointestinal tube, skin epidermis, etc.) and at the inner surface of cavities in other organs (i.e., exocrine system in the liver and pancreas, collective tubes in the kidney's nephrons, blood and lymphoid vessels, etc.).

Eukaryotes Class of unicellular or multicellular organisms in which the cells contain membrane-limited nucleus and membrane-limited organelles including endoplasmic reticulum (ER), mitochondrion (animal and fungi), and chloroplast (vegetal), which constitutes one of the three distinct evolutionary lineages of modern-day organisms; also called *Eukarya*.

Extracellular matrix (ECM) Insoluble network constituted by fibrous proteins, adhesive proteins, and polysaccharides secreted by most animal cells.

Fertilization Fusion of a female and male haploid gamete to form a diploid zygote.

Fibroblast A particular cell type present in connective tissues.

Gamete A cell type produced after meiosis by germ cells carrying half of the genetic material (haploid), currently named sperm (males) and oocytes or eggs (females).

Gastrula Early embryonic stage after the blastula which undergoes cell differentiation to produce three embryonic layers (endoderm, mesoderm, and ectoderm) and a process of invagination of the tissues ends in the formation of the primitive gastrointestinal tube (gut).

Gene It is basically a DNA sequence present in chromosomes necessary to produce an RNA or protein molecules.

Generation After fecundation and zygote formation, is the period of time needed for a particular organism to produce offspring.

Genetic drift Describes random fluctuations in the number of gene variants in a determined population.

Genome Total genetic information carried by a cell or organism.

Germ cell Any precursor cell that can give rise to gametes.

Germline Lineage of germ cells that can give rise to gametes.

Gonads Sexual organs in female (ovaries) and male (testicles).

Haploid An organism or a cell carrying only one of the two homologous chromosomes (in the case of a diploid organism or cell). Gametes and bacterial cells are haploid.

Heterotroph A group of organisms that eats other organisms (animal, plant, or bacteria) to produce energy.

Histones Histones are basic proteins with barrel-like structure used to roll up the DNA molecule to minimize its volume and fit into the cell nucleus forming the essential component of the eukaryotic chromosomes: the chromatin.

Homologous chromosome One of the two copies of each type of chromosome present in a diploid cell, called homologue. Moreover, each homologue is derived from a different parent.

Inner cell mass (ICM) The internal cells of the blastocyst that will contribute to all the tissues and organs of the embryo.

Lichen An organism formed by the symbiosis (coexistence) of and algae and a cyanobacteria or fungi.

Meiosis A special type of cellular division in eukaryotes taking place during germ cell maturation to produce gametes. Consisting of one round of DNA replication followed by two rounds of nuclear and cellular divisions where the initial diploid cell ends producing four haploid cells (gametes).

Mesoderm An embryonic cell layer present in between the ectoderm and the endoderm which gives rise to many tissues including the bone, cartilage, fat, muscle, blood, skin dermis, and others.

Metabolism Consist of all biochemical reactions that take place within a cell which provides the catabolism, the anabolism, and the necessary energy for subsistence.

Metamorphosis Drastic transformation that occurs during development of certain animals such as fly from larva, froglets from tadpoles, or butterflies from chrysalids.

Mitochondrion (plural: mitochondria) A membrane-limited organelle present in eukaryotic cells which contains its own circular DNA and is in charge of producing most of the ATP during respiration.

Mitosis The process in eukaryotic cells consisting in replicating their genetic material (DNA) present in the nucleus to form two nuclei that are segregated to each of the two daughter cells.

Mutation An irreversible change that could happen in the structure of the DNA molecule.

Multicellular organism Are organisms that consist of more than one cell.

Niche Describes as the particular environment surrounding a cell, organisms of group of organism.

Nucleic acid A polymer of nucleotides linked by phosphodiester bonds, including DNA and RNA.

Nucleus A membrane-limited organelle in eukaryotic cells containing the DNA in the form of chromosomes.

Oocyte Developing egg cell.

Organelle A membrane-surrounded structure present in the cytoplasm of eukaryotic cells.

Origin of life Refers to the process by which life has emerged from nonliving matter, including organic and inorganic matter and its possible combinations.

Phagocytosis Process by which particles are engulfed by eukaryotic cells.

Phenotype The way a gene activity is expressed by manifesting in an organism.

Photosynthesis Is the process by which plants, algae, and cyanobacteria capture photons (light energy) and used to convert carbon dioxide and water into oxygen and high-energy organic compounds (like ATP).

Phototropic An organism that uses light energy to produce complex organic compounds such as sugars and acquires energy.

Placenta Organ formed by maternal and embryonic tissue that serves to nurturing mammalian embryos.

Plasma membrane The structure surrounding cells that separate them from its external environment, formed by a phospholipid bilayer and associated proteins.

Polysaccharide Linear or branched polymer of monosaccharides.

Prokaryotes Eubacteria and archaea that lack nucleus and other organelles.

Quiescent A cell that is out of the cell cycle and is in the G0 state.

Recombination A process in which homologous chromosomes exchange fragments to give new combinations.

Respiration A cellular process involving the uptake of O_2 coupled to production of CO_2.

Reproduction The production of offspring by a sexual or asexual process.

Ribonucleic acid (RNA) Linear single-stranded polymer formed of ribose nucleotides.

Somatic cell Any plant or animal cell other than a germ cell, germline, or germ stem cell.

Stem cell A self-renewing cell that divides to give rise to a cell with an identical differentiation potential and another called progenitor, which normally will undergo differentiation into a specific tissue cell type.

Syncytium Cells containing many nuclei particularly present during the first stages of fly development.

Telomere End region present at the end of linear eukaryotic chromosomes containing characteristic telomeric (TEL) sequences.

Tissue regeneration Natural process constantly undergoing in tissues that consist in replacing death cells by new ones involving cell division. Also called physiological regeneration or tissue homeostasis. It can be also used to describe the process that tissues undergo after injury (also described as pathological regeneration), only if identical tissue structure and function are obtained after the process, otherwise is called tissue repair.

Tissue repair Natural process that tissues undergo after injury if partial or nonidentical tissue structure and/or function is obtained after the process, which normally ends in scar formation.

Tumor Tissue abnormalities normally caused by specific mutations in key genes.

Unicellular organism Are organisms that consist of only one cell such as bacteria, archaebacteria, amoeba, and protozoa.

Virus A very tiny cell parasite consisted of well-coated nucleic acid—in the form of RNA or DNA—which requires a host cell to replicate as well as propagate.

Zygote A fertilized egg resulting from the fusion of a male and a female gamete.

Index[1]

[1] Note: Page numbers followed by 'n' refer to notes.

Printed in the United States
by Baker & Taylor Publisher Services